Stochastic Processes
An Introduction
Third Edition

CHAPMAN & HALL/CRC
Texts in Statistical Science Series

Series Editors
Joseph K. Blitzstein, *Harvard University, USA*
Julian J. Faraway, *University of Bath, UK*
Martin Tanner, *Northwestern University, USA*
Jim Zidek, *University of British Columbia, Canada*

Extending the Linear Model with R: Generalized Linear, Mixed Effects and Nonparametric Regression Models, Second Edition
J.J. Faraway

Linear Models with R, Second Edition
J.J. Faraway

A Course in Large Sample Theory
T.S. Ferguson

Multivariate Statistics: A Practical Approach
B. Flury and H. Riedwyl

Readings in Decision Analysis
S. French

Discrete Data Analysis with R: Visualization and Modeling Techniques for Categorical and Count Data
M. Friendly and D. Meyer

Markov Chain Monte Carlo: Stochastic Simulation for Bayesian Inference, Second Edition
D. Gamerman and H.F. Lopes

Bayesian Data Analysis, Third Edition
A. Gelman, J.B. Carlin, H.S. Stern, D.B. Dunson, A. Vehtari, and D.B. Rubin

Multivariate Analysis of Variance and Repeated Measures: A Practical Approach for Behavioural Scientists
D.J. Hand and C.C. Taylor

Practical Longitudinal Data Analysis
D.J. Hand and M. Crowder

Logistic Regression Models
J.M. Hilbe

Richly Parameterized Linear Models: Additive, Time Series, and Spatial Models Using Random Effects
J.S. Hodges

Statistics for Epidemiology
N.P. Jewell

Stochastic Processes: An Introduction, Third Edition
P.W. Jones and P. Smith

The Theory of Linear Models
B. Jørgensen

Pragmatics of Uncertainty
J.B. Kadane

Principles of Uncertainty
J.B. Kadane

Graphics for Statistics and Data Analysis with R
K.J. Keen

Mathematical Statistics
K. Knight

Introduction to Functional Data Analysis
P. Kokoszka and M. Reimherr

Introduction to Multivariate Analysis: Linear and Nonlinear Modeling
S. Konishi

Nonparametric Methods in Statistics with SAS Applications
O. Korosteleva

Modeling and Analysis of Stochastic Systems, Third Edition
V.G. Kulkarni

Exercises and Solutions in Biostatistical Theory
L.L. Kupper, B.H. Neelon, and S.M. O'Brien

Exercises and Solutions in Statistical Theory
L.L. Kupper, B.H. Neelon, and S.M. O'Brien

Design and Analysis of Experiments with R
J. Lawson

Design and Analysis of Experiments with SAS
J. Lawson

A Course in Categorical Data Analysis
T. Leonard

Statistics for Accountants
S. Letchford

Introduction to the Theory of Statistical Inference
H. Liero and S. Zwanzig

Statistical Theory, Fourth Edition
B.W. Lindgren

Stationary Stochastic Processes: Theory and Applications
G. Lindgren

Statistics for Finance
E. Lindström, H. Madsen, and J. N. Nielsen

The BUGS Book: A Practical Introduction to Bayesian Analysis
D. Lunn, C. Jackson, N. Best, A. Thomas, and D. Spiegelhalter

Introduction to General and Generalized Linear Models
H. Madsen and P. Thyregod

Time Series Analysis
H. Madsen

Pólya Urn Models
H. Mahmoud

Texts in Statistical Science

Stochastic Processes
An Introduction
Third Edition

Peter W. Jones
Peter Smith

CRC Press
Taylor & Francis Group
Boca Raton London New York

CRC Press is an imprint of the
Taylor & Francis Group, an **informa** business
A CHAPMAN & HALL BOOK

CRC Press
Taylor & Francis Group
6000 Broken Sound Parkway NW, Suite 300
Boca Raton, FL 33487-2742

First issued in paperback 2020

ISBN-13: 978-1-4987-7811-4 (hbk)
ISBN-13: 978-0-367-65760-4 (pbk)

Library of Congress Cataloging-in-Publication Data

Names: Jones, P. W. (Peter Watts), 1945- | Smith, Peter, 1935-.
Title: Stochastic processes : an introduction / Peter W. Jones & Peter Smith.
Description: Third edition. | Boca Raton : Chapman & Hall/CRC Press, 2017. |
Series: Chapman & Hall/CRC Texts in Statistical Science series | Includes bibliographical
    references and index.
Identifiers: LCCN 2017008318| ISBN 9781498778114 (hardback) | ISBN 9781315156576
    (e-book) | ISBN 9781498778121 (adobe reader) | ISBN 9781498779234 (epub) | ISBN
    9781351633246 (mobi/kindle).
Subjects: LCSH: Stochastic processes--Textbooks.
Classification: LCC QA274 .J66 2017 | DDC 519.2/3--dc23
LC record available at https://lccn.loc.gov/2017008318

Visit the Taylor & Francis Web site at
http://www.taylorandfrancis.com

and the CRC Press Web site at
http://www.crcpress.com

Contents

Preface to the Third Edition

This textbook was developed from a course in stochastic processes given by the authors over many years to second-year students studying mathematics or statistics at Keele University. At Keele the majority of students take degrees in mathematics or statistics jointly with another subject, which may be from the sciences, social sciences, or humanities. For this reason the course has been constructed to appeal to students with varied academic interests, and this is reflected in the book by including applications and examples that students can quickly understand and relate to. In particular, in the earlier chapters, the classical gambler's ruin problem and its variants are modeled in a number of ways to illustrate simple random processes. Specialized applications have been avoided to accord with our view that students have enough to contend with in the mathematics required in stochastic processes.

The book is to a large extent modular, and topics can be selected from Chapters 2 to 10 for a one-semester course in random processes. It is assumed that readers have already encountered the usual first-year courses in calculus and matrix algebra and have taken a first course in probability; nevertheless, a revision of relevant basic probability is included for reference in Chapter 1. Some of the easier material on discrete random processes is included in Chapters 2, 3, and 4, which cover some simple gambling problems, random walks, and Markov chains. Random processes continuous in time are developed in Chapters 5 and 6. These include birth and death processes, and general population models. Continuous time models include queues in Chapter 7, which has an extended discussion on the analysis of associated stationary processes. There follow two chapters on reliability and other random processes, the latter including branching processes and martingales. The main text ends with a new chapter on Brownian motion, which we have attempted to explain in an intuitive manner.

Much of the text has been extensively reworked to clarify explanations. New problems and worked examples have been added in this edition. Some applications have been added including wildlife and stochastic epidemic models. There are over 50 worked examples in the text and 220 end-of-chapter problems with hints and answers listed at the end of the book. An Appendix of key mathematical terms is included for reference.

The software $Mathematica^{TM}$ has been used to evaluate many numerical examples and simulations, and to check results symbolically. Many of the graphs and figures have been produced using this software.

An alternative software for some applications is *R*, which is a statistical computing and graphics package available free of charge; it can be downloaded from:

http://www.r-project.org

Like *R*, *S-PLUS* (not freeware) is derived from the *S* language, and hence users of these packages will be able to apply them to the solution of numerical projects, including those involving matrix algebra presented in the text. *Mathematica* code has been applied to all the projects listed (about 40 in total) by chapters in Chapter 11, and *R* code to some as appropriate.

The programs, which generally use standard commands, are intended to be flexible in that inputs, parameters, data, etc., can be varied by the user. Graphs and computations can often add insight into what might otherwise be viewed as rather mechanical analysis. In addition, more complicated examples, which might be beyond hand calculations, can be attempted. However, the main text does not assume or require any particular software.

A *Solutions Manual* and *Mathematica and R Programs* are freely available at:

www.crcpress.com/9781498778114

The website contains solutions to all end-of-chapter problems and all *Mathematica* and *R* programs listed in Chapter 11. They should be considered as an extension of the main text.

We should acknowledge the influence of the internet. Any individual studying stochastic processes (or any other mathematical course for that matter) would do well to search topics on the web for alternative viewpoints and code for other software. The quality can vary but there are also many excellent lecture notes, actual lectures, presentations, video graphics and applications.

Finally, we would like to thank the many students at Keele over many years who have helped to develop this book, and acknowledge the interest shown by users of the first and second editions in helping us to refine and update this new edition. We are particularly grateful to a number of reviewers who have made frank and detailed comments, and suggestions for improvement in this new edition.

Peter W. Jones
Peter Smith
Keele University
2017

CHAPTER 1

Some Background on Probability

1.1 Introduction

We shall be concerned with the modeling and analysis of random experiments using the theory of probability. The outcome of such an experiment is the result of a
stochastic or *random process*. In particular we shall be interested in the way in which
the results or outcomes vary or evolve over time. An experiment or trial is any situation where an outcome is observed. In many of the applications considered, these
outcomes will be numerical, sometimes in the form of counts or enumerations. The
experiment is random if the outcome is not predictable or is uncertain.

At first we are going to be concerned with simple mechanisms for creating random
outcomes, namely games of chance. One recurring theme initially will be the study of
the classical problem known as *gambler's ruin*. We will then move on to applications
of probability to modeling in, for example, engineering, medicine, and biology. We
make the assumption that the reader is familiar with the basic theory of probability.
This background will however be reinforced by the brief review of these concepts
which will form the main part of this chapter.

1.2 Probability

In random experiments, the list of all possible outcomes is termed the **sample space**,
denoted by S. This list consists of individual **outcomes** or **elements**. Sample spaces
can have a finite or infinite number of outcomes, and can be discrete or continuous.
These elements have the properties that they are **mutually exclusive** and that they
are **exhaustive**. Mutually exclusive means that two or more outcomes cannot occur
simultaneously; exhaustive means that *all* possible outcomes are in the list. Thus each
time the experiment is carried out one of the outcomes in S must occur. A collection
of elements of S is called an **event**: these are usually denoted by capital letters,
A, B, etc. We denote by $\mathbf{P}(A)$ the **probability** that the event A will occur at each
repetition of the random experiment. Remember that A is said to have occurred if one
element making up A has occurred. In order to calculate or estimate the probability
of an event A there are two possibilities. In one approach an experiment can be
performed a large number of times, and $\mathbf{P}(A)$ can be approximated by the relative
frequency with which A occurs. In order to analyse random experiments we make
the assumption that the conditions surrounding the trials remain the same, and are
independent of one another. We hope that some regularity or settling down of the

outcome is apparent. The ratio

$$\frac{\text{the number of times a particular event } A \text{ occurs}}{\text{total number of trials}}$$

is known as the **relative frequency** of the event, and the number to which it appears to converge as the number of trials increases is known as the probability of an outcome within A. Where we have a finite sample space it might be reasonable to assume that the outcomes of an experiment are equally likely to occur as in the case, for example, in rolling a fair die or spinning an unbiased coin. In this case the probability of A is given by

$$\mathbf{P}(A) = \frac{\text{number of elements of } S \text{ where } A \text{ occurs}}{\text{number of elements in } S}.$$

There are, of course, many 'experiments' which are not repeatable. Horse races are only run once, and the probability of a particular horse winning a particular race may not be calculated by relative frequency. However, a punter may form a view about the horse based on other factors which may be repeated over a series of races. The past form of the horse, the form of other horses in the race, the state of the course, the record of the jockey, etc., may all be taken into account in determining the punter's estimate of the probability of a win. This leads to a view of probability as a *'degree of belief'* about uncertain outcomes. The odds placed by bookmakers on the horses in a race reflect how punters place their bets on the race. The odds are also set so that the bookmakers expect to make a profit.

It is convenient to use **set** notation when deriving probabilities of events. This leads to the sample space S being termed the **universal set**, the set of all outcomes: an event A is a **subset** of S. This also helps with the construction of more complex events in terms of the unions and intersections of several events. The **Venn diagrams**[1] shown in Figure 1.1 represent the main set operations of **union** ($\cup$), **intersection** ($\cap$), and **complement** (A^c) which are required in probability.

- **Union**. The union of two sets A and B is the set of all elements which belong to A, or to B, or to both. It can be written formally as

$$A \cup B = \{x | x \in A \text{ or } x \in B \text{ or both}\}.$$

- **Intersection**. The intersection of two sets A and B is the set $A \cap B$ which contains all elements common to both A and B. It can be written as

$$A \cap B = \{x | x \in A \text{ and } x \in B\}.$$

- **Complement**. The complement A^c of a set A is the set of all elements which belong to the universal set S but do not belong to A. It can be written as

$$A^c = \{x \notin A\}.$$

So, for example, in an experiment in which we are interested in two events A and B, then $A^c \cap B$ may be interpreted as 'only B', being the intersection of the **complement** of A and B (see Figure 1.1d): this is alternatively expressed in the

[1] John Venn (1834–1923), English philosopher.

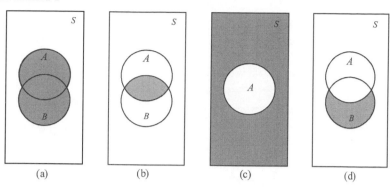

Figure 1.1 *(a) The union $A \cup B$ of A and B; (b) the intersection $A \cap B$ of A and B; (c) the complement A^c of A: S is the universal set; (d) $A^c \cap B$ or $B \backslash A$.*

difference notation $B \backslash A$ meaning B but not A. We denote by ϕ the **empty set**, that is, the set which contains no elements. Note that $S^c = \phi$. Two events A and B are said to be **mutually exclusive** if A and B have no events in common so that $A \cap B = \phi$, the empty set: in set terminology A and B are said to be **disjoint** sets.

The probability of any event satisfies, the three axioms

- **Axiom 1**: $0 \leq \mathbf{P}(A) \leq 1$ for every event A
- **Axiom 2**: $\mathbf{P}(S) = 1$
- **Axiom 3**: $\mathbf{P}(A \cup B) = \mathbf{P}(A) + \mathbf{P}(B)$ if A and B are mutually exclusive ($A \cap B = \phi$)

Axiom 3 may be extended to more than two mutually exclusive events, say k of them represented by

$$A_1, A_2, \ldots, A_k$$

in S where $A_i \cap A_j = \phi$ for all $i \neq j$. This is called a **partition** of S if

- (a) $A_i \cap A_j = \phi$ for all $i \neq j$,
- (b) $\bigcup_{i=1}^{k} A_i = A_1 \cup A_2 \cup \ldots \cup A_k = S$: $A_1, A_2, \ldots A_k$ is an **exhaustive** list so that one of the events must occur.
- (c) $\mathbf{P}(A_i) > 0$.

In this definition, (a) states that the events are mutually exclusive, (b) that every event in S occurs in one of the events A_i, and (c) implies that there is a nonzero probability that any A_i occurs. It follows that

$$1 = \mathbf{P}(S) = \mathbf{P}(A_1 \cup A_2 \cup \cdots \cup A_k) = \sum_{i=1}^{k} \mathbf{P}(A_i).$$

Theorem

- (a) $\mathbf{P}(A^c) = 1 - \mathbf{P}(A)$;

- (b) $\mathbf{P}(A \cup B) = \mathbf{P}(A) + \mathbf{P}(B) - \mathbf{P}(A \cap B)$.

(a) Axiom 3 may be combined with Axiom 2 to give $\mathbf{P}(A^c)$, the probability that the complement A^c occurs, by noting that $S = A \cup A^c$. This is a partition of S into the mutually exclusive exhaustive events A and A^c. Thus

$$1 = \mathbf{P}(S) = \mathbf{P}(A \cup A^c) = \mathbf{P}(A) + \mathbf{P}(A^c),$$

giving

$$\mathbf{P}(A^c) = 1 - \mathbf{P}(A).$$

(b) For any sets A and B

$$A \cup B = A \cup (B \cap A^c),$$

and

$$B = (A \cap B) \cup (B \cap A^c),$$

in which A and $B \cap A^c$ are disjoint sets, and $A \cap B$ and $B \cap A^c$ are disjoint sets. Therefore, by Axiom 3,

$$\mathbf{P}(A \cup B) = \mathbf{P}(A) + \mathbf{P}(B \cap A^c),$$

and

$$\mathbf{P}(B) = \mathbf{P}(A \cap B) + \mathbf{P}(B \cap A^c).$$

Elimination of $\mathbf{P}(B \cap A^c)$ between these equations leads to

$$\mathbf{P}(A \cup B) = \mathbf{P}(A) + \mathbf{P}(B) - \mathbf{P}(A \cap B) \tag{1.1}$$

as required.

Example 1.1. *Two distinguishable fair dice a and b are rolled and the values on the uppermost faces noted. What are the elements of the sample space? What is the probability that the sum of the face values of the two dice is 7? What is the probability that at least one 5 appears?*

We distinguish first the outcome of each die so that there are $6 \times 6 = 36$ possible outcomes for the pair. The sample space has 36 elements of the form (i, j) where i and j take all integer values $1, 2, 3, 4, 5, 6$, and i is the outcome of die a and j is the outcome of b. The full list is

$$
\begin{aligned}
S = \{ \quad & (1,1), (1,2), (1,3), (1,4), (1,5), (1,6), \\
& (2,1), (2,2), (2,3), (2,4), (2,5), (2,6), \\
& (3,1), (3,2), (3,3), (3,4), (3,5), (3,6), \\
& (4,1), (4,2), (4,3), (4,4), (4,5), (4,6), \\
& (5,1), (5,2), (5,3), (5,4), (5,5), (5,6), \\
& (6,1), (6,2), (6,3), (6,4), (6,5), (6,6) \quad \},
\end{aligned}
$$

and they are all assumed to be equally likely since the dice are fair. If A_1 is the event that the sum of the dice is 7, then from the list,

$$A_1 = \{(1,6), (2,5), (3,4), (4,3), (5,2), (6,1)\}$$

which occurs for 6 elements out of 36. Hence

$$P(A_1) = \tfrac{6}{36} = \tfrac{1}{6}.$$

The event that at least one 5 appears is the list

$$A_2 = \{(1,5), (2,5), (3,5), (4,5), (5,1), (5,2), (5,3), (5,4), (5,5), (5,6), (6,5)\},$$

which has 11 elements. Hence
$$P(A_2) = \tfrac{11}{36}.$$

Example 1.2 *From a well-shuffled pack of 52 playing cards a single card is randomly drawn. Find the probability that it is a heart or an ace.*

Let A be the event that the card is an ace, and B the event that it is a heart. The event $A \cap B$ is the ace of hearts. We require the probability that it is an ace or a heart, which is $\mathbf{P}(A \cup B)$. However, since one of the aces is a heart the events are not mutually exclusive. Hence, we must use Eqn (1.25). It follows that

the probability that an ace is drawn is $\mathbf{P}(A) = 4/52$,

the probability that a heart is drawn is $\mathbf{P}(B) = 13/52 = 1/4$,

the probability that the ace of hearts is drawn is $\mathbf{P}(A \cap B) = 1/52$.

From (1.25)
$$\mathbf{P}(A \cup B) = \mathbf{P}(A) + \mathbf{P}(B) - \mathbf{P}(A \cap B) = \frac{4}{52} + \frac{1}{4} - \frac{1}{52} = \frac{16}{52} = \frac{4}{13}.$$

This example illustrates events which are not mutually exclusive. The result could also be obtained directly by noting that 16 of the 52 cards are either hearts or aces.

In passing note that $A \cap B^c$ is the set of aces excluding the ace of hearts, whilst $A^c \cap B$ is the heart suit excluding the ace of hearts. Hence
$$\mathbf{P}(A \cap B^c) = \frac{3}{52}, \qquad \mathbf{P}(A^c \cap B) = \frac{12}{52} = \frac{3}{13}.$$

1.3 Conditional probability and independence

If the occurrence of an event B is affected by the occurrence of another event A then we say that A and B are dependent events. We might be interested in a random experiment with which A and B are associated. When the experiment is performed, it is known that event A has occurred. Does this affect the probability of B? This probability of B now becomes the **conditional probability** of B given A, which is now written as $\mathbf{P}(B|A)$. Usually this will be distinct from the probability $\mathbf{P}(B)$. Strictly speaking, this probability is conditional since we must assume that B is conditional on the sample space occurring, but it is implicit in $\mathbf{P}(B)$. On the other hand the conditional probability of B is restricted to that part of the sample space where A has occurred. This conditional probability is defined as

$$\mathbf{P}(B|A) = \frac{\mathbf{P}(A \cap B)}{\mathbf{P}(A)}, \qquad \mathbf{P}(A) > 0. \tag{1.2}$$

In terms of counting, suppose that an experiment is repeated N times, of which A occurs $\mathrm{N}(A)$ times, and A given by B occurs $\mathrm{N}(B \cap A)$ times. The proportion of times that B occurs is

$$\frac{\mathrm{N}(A \cap B)}{\mathrm{N}(A)} = \frac{\mathrm{N}(A \cap B)}{\mathrm{N}} \frac{\mathrm{N}}{\mathrm{N}(A)},$$

which supports (1.2).

If the probability of B is unaffected by the prior occurrence of A, then we say that A and B are **independent** or that

$$\mathbf{P}(B|A) = \mathbf{P}(B),$$

which from the above implies that

$$\mathbf{P}(A \cap B) = \mathbf{P}(A)\mathbf{P}(B).$$

Conversely, if $\mathbf{P}(B|A) = \mathbf{P}(B)$, then A and B are independent events. Again this result can be extended to 3 or more independent events.

Example 1.3 *Let A and B be independent events with $\mathbf{P}(A) = \frac{1}{4}$ and $\mathbf{P}(B) = \frac{2}{3}$. Calculate the following probabilities: (a) $\mathbf{P}(A \cap B)$; (b) $\mathbf{P}(A \cap B^c)$; (c) $\mathbf{P}(A^c \cap B^c)$; (d) $\mathbf{P}(A^c \cap B)$; (e) $\mathbf{P}((A \cup B)^c)$.*

Since the events are independent, then $\mathbf{P}(A \cap B) = \mathbf{P}(A)\mathbf{P}(B)$. Hence
(a) $\mathbf{P}(A \cap B) = \frac{1}{4} \cdot \frac{2}{3} = \frac{1}{6}$.
(b) The independence A and B^c follows by eliminating $\mathbf{P}(A \cap B)$ between the equations

$$\mathbf{P}(A \cap B) = \mathbf{P}(A)\mathbf{P}(B) = \mathbf{P}(A)[1 - \mathbf{P}(B^c)]$$

and

$$\mathbf{P}(A) = \mathbf{P}[(A \cap B^c) \cup (A \cap B)] = \mathbf{P}(A \cap B^c) + \mathbf{P}(A \cap B).$$

Hence

$$\mathbf{P}(A \cap B^c) = \mathbf{P}(A)\mathbf{P}(B^c) = \mathbf{P}(A)[1 - \mathbf{P}(B)] = \frac{1}{4}(1 - \frac{2}{3}) = \frac{1}{12}.$$

(c) Since A^c and B^c are independent events,

$$\mathbf{P}(A^c \cap B^c) = \mathbf{P}(A^c)\mathbf{P}(B^c) = [1 - \mathbf{P}(A)][1 - \mathbf{P}(B)] = \frac{3}{4} \cdot \frac{1}{3} = \frac{1}{4}.$$

(d) Since A^c and B are independent events, $\mathbf{P}(A^c \cap B) = \mathbf{P}(A^c)\mathbf{P}(B) = [1 - \frac{1}{4}]\frac{2}{3} = \frac{1}{2}$.
(e) $\mathbf{P}((A \cup B)^c) = 1 - \mathbf{P}(A \cup B) = 1 - \mathbf{P}(A) - \mathbf{P}(B) + \mathbf{P}(A \cap B)$ by (1.1). Hence

$$\mathbf{P}((A \cup B)^c) = 1 - \mathbf{P}(A) - \mathbf{P}(B) + \mathbf{P}(A)\mathbf{P}(B) = 1 - \frac{1}{4} - \frac{2}{3} + \frac{1}{6} = \frac{1}{4}.$$

Example 1.4. *For three events A, B, and C, show that*

$$\mathbf{P}(A \cap B|C) = \mathbf{P}(A|B \cap C)\mathbf{P}(B|C),$$

where $\mathbf{P}(C) > 0$.

By using (1.2) and viewing $A \cap B \cap C$ as $(A \cap B) \cap C$ or $A \cap (B \cap C)$,

$$\mathbf{P}(A \cap B \cap C) = \mathbf{P}(A \cap B|C)\mathbf{P}(C) = \mathbf{P}(A|B \cap C)\mathbf{P}(B \cap C).$$

Hence

$$\mathbf{P}(A \cap B|C) = \mathbf{P}(A|B \cap C)\frac{\mathbf{P}(B \cap C)}{\mathbf{P}(C)} = \mathbf{P}(A|B \cap C)\mathbf{P}(B|C)$$

by (1.2) again.

Example 1.5. *The security lock on a case can be opened by entering 2 digits (each from $1, 2, \ldots, 9$) in the lock. which will have $10^2 = 100$ possible codes. A traveller has forgotten the code, and attempts to find the code by choosing 2 digits at random. If the code fails to open* *the case, the traveller tries another code from the remaining pairs. What is the probability that the new code opens the case, or any subsequent attempt?*

Let X_n be the event that the n-th attempt opens the case, previous attempts having failed. At the first attempt $\mathbf{P}(X_1) = 1/100$. Now

$$\mathbf{P}(X_2) = \mathbf{P}(X_2 \cap X_1^c),$$

where complement X_1^c is the event that the first attempt failed. Using (1.2)

$$\mathbf{P}(X_2) = \mathbf{P}(X_2 \cap X_1^c) = \mathbf{P}(X_1^c)\mathbf{P}(X_2|X_1^c) = \frac{9}{100} \cdot \frac{1}{99} = \frac{1}{100},$$

the same probability as the first attempt. In fact $\mathbf{P}(X_n) = 1/100$ for all n. Note that the *conditional probability* $\mathbf{P}(X_2|X_1^c) = 1/99$.

A result known as the **law of total probability** or the **partition theorem** will be used extensively later, for example, in the discrete gambler's ruin problem (Section 2.1) and the Poisson [2] process (Section 5.2). Suppose that $A_1, A_2, \ldots, A_k$ represents a partition of S into k mutually exclusive, exhaustive events in which, interpreted as sets, the sets fill the space S but with none of the sets overlapping. Figure 1.2 shows such a scheme. When a random experiment takes place, one and only

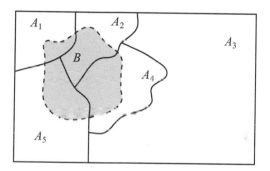

Figure 1.2 *Schematic set view of a partition of S into 5 events $A_1, \ldots, A_5$ with event B intersecting all 5 events.*

one of the events can take place.

Suppose that B is another event associated with the same random experiment (Figure 1.2). Then B must be made up of the sum of the intersections of B with events in the partition. Some of these will be empty but this does not matter. We can say that B is the union of the intersections of B with each A_i. Thus

$$B = \bigcup_{i=1}^{k} B \cap A_i,$$

but the significant point is that any pair of these events is mutually exclusive. It

---

[2] Siméon-Denis Poisson (1781–1840), French mathematician.

follows that

$$\mathbf{P}(B) = \sum_{i=1}^{k} \mathbf{P}(B \cap A_i). \tag{1.3}$$

Since, from Eqn (1.2),

$$\mathbf{P}(B \cap A_i) = \mathbf{P}(B|A_i)\mathbf{P}(A_i),$$

Eqn (1.3) can be expressed as

$$\mathbf{P}(B) = \sum_{i=1}^{k} \mathbf{P}(B|A_i)\mathbf{P}(A_i), \tag{1.4}$$

which is the law of total probability or the partition theorem.

1.4 Discrete random variables

In most of the applications considered in this text, the outcome of the experiment will be numerical. A **random variable** usually denoted by the capital letters X, Y, or Z, say, is a numerical value associated with the outcome of a random experiment. If s is an element of the original sample space S, which may be numerical or symbolic, then $X(s)$ is a real number associated with s. The same experiment, of course, may generate several random variables. Each of these random variables will, in turn, have sample spaces whose elements are usually denoted by lower case letters such as $x_1, x_2, x_3, \ldots$ for the random variable X. We are now interested in assigning probabilities to events such as $\mathbf{P}(X = x_1)$, the probability that the random variable X is x_1, or $\mathbf{P}(X \leq x_2)$, the probability that the random variable is less than or equal to x_2.

If the sample space is finite or countably infinite on the integers (that is, the elements $x_0, x_1, x_2, \ldots$ can be counted against integers, say $0, 1, 2, \ldots$) then we say that the random variable is **discrete**. Technically, the set $\{x_i\}$ will be a countable subset $\mathcal{V}$, say, of the real numbers $\mathcal{R}$. We can represent the $\{x_i\}$ generically by the variable x with $x \in \mathcal{V}$. For example, $\mathcal{V}$ could be the set

$$\{0, \tfrac{1}{2}, 1, \tfrac{3}{2}, 2, \tfrac{5}{2}, 3, \ldots\}.$$

In many cases $\mathcal{V}$ consists simply of the integers or a subset of the integers, such as

$$\mathcal{V} = \{0, 1\} \quad \text{or} \quad \mathcal{V} = \{0, 1, 2, 3, \ldots\}.$$

In the random walks of Chapter 3, however, $\mathcal{V}$ may contain all the positive and negative integers

$$\ldots -3, -2, -1, 0, 1, 2, 3, \ldots.$$

In these *integer* cases we can put $x_i = i$.

The probability denoted by

$$p(x_i) = \mathbf{P}(X = x_i)$$

is known as the **probability mass function**. The pairs $\{x_i, p(x_i)\}$ for all i in the sample space define the **probability distribution** of the random variable X. If $x_i =$

i, which occurs frequently in applications, then $p(x_i) = p(i)$ is often replaced by p_i. Since the x values are mutually exclusive and exhaustive then it follows that for $p(x_0), p(x_1), \ldots$

- (a) $0 \le p(x_i) \le 1$ for all i,

- (b) $\sum_{i=0}^{\infty} p(x_i) = 1$, or in generic form $\sum_{x \in V} p(x) = 1$,

- (c) $\mathbf{P}(X \le x_k) = \sum_{i=0}^{k} p(x_i)$, which is known as the **distribution function** for $k = 0, 1, 2, \ldots$.

Example 1.5. *A fair die is rolled until the first 6 appears face up. Find the probability that the first 6 appears at the n-th throw.*

Let the random variable N be the number of throws until the first 6 appears face up. This is an example of a discrete random variable N with an infinite number of possible outcomes

$$\{1, 2, 3, \ldots\}.$$

The probability of a 6 appearing for any throw is $\frac{1}{6}$ and of any other number appearing is $\frac{5}{6}$. Hence the probability of $n - 1$ numbers other than 6 appearing followed by a 6 is

$$\mathbf{P}(N = n) = \left(\frac{5}{6}\right)^{n-1}\left(\frac{1}{6}\right) = \frac{5^{n-1}}{6^n},$$

which is the probability mass function for this random variable.

In this example the distribution has the probability

$$\mathbf{P}(N \le k) = \frac{1}{6} + \frac{1}{6}\frac{5}{6} + \cdots + \frac{1}{6}\left(\frac{5}{6}\right)^{k-1} = \frac{1}{6}\sum_{i=1}^{k}\left(\frac{5}{6}\right)^{i-1} = 1 - \left(\frac{5}{6}\right)^k,$$

for $k = 1, 2, \ldots 6$ after summing the geometric series.

1.5 Continuous random variables

In many applications the discrete random variable, which for example might take the integer values $1, 2, \ldots$, is inappropriate for problems where the random variable can take any real value in an interval. For example, the random variable T could be the time measured from time $t = 0$ until a light bulb fails. This could be any value $t \ge 0$. In this case T is called a **continuous random variable**. Generally, if X is a continuous random variable there are mathematical difficulties in defining the event $X = x$: the probability is usually defined to be zero. Probabilities for continuous random variables may only be defined over intervals of values as, for example, in $\mathbf{P}(x_1 < X < x_2)$.

We define a **probability density function** (pdf) $f(x)$ over $-\infty < x < \infty$, which has the properties:

- (a) $f(x) \ge 0, (-\infty < x < \infty)$;

- (b) $\mathbf{P}(x_1 \leq X \leq x_2) = \int_{x_1}^{x_2} f(x)dx$ for any x_1, x_2 such that $-\infty < x_1 < x_2 < \infty$;

- (c) $\int_{-\infty}^{\infty} f(x)dx = 1$.

A possible graph of a density function $f(x)$ is shown in Figure 1.3. By (a) the area the curve must remain nonnegative, by (b) the probability that X lies between x_1

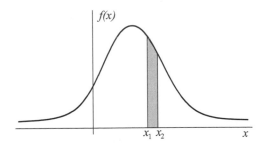

Figure 1.3 *A probability density function.*

and x_2 is the shaded area, and by (c) the total area under the curve must be 1 since $\mathbf{P}(-\infty < X < \infty) = 1$.

We define the **(cumulative) distribution function** (cdf) $F(x)$ as the probability that X is less than or equal to x. Thus

$$F(x) = \mathbf{P}(X \leq x) = \int_{-\infty}^{x} f(u)du. \tag{1.5}$$

It follows from (c) above that

$$F(x) \to 1 \quad \text{as} \quad x \to \infty,$$

and that

$$\mathbf{P}(x_1 \leq x \leq x_2) = \int_{x_1}^{x_2} f(u)du = F(x_2) - F(x_1). \tag{1.6}$$

An example of a cdf is shown in Figure 1.4.

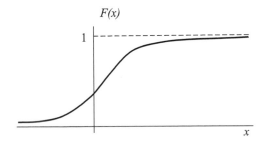

Figure 1.4 *A (cumulative) distribution function.*

Example 1.6. *Show that*

$$f(x) = \begin{cases} 1/(b-a) & a \le x \le b \\ 0 & \text{for all other values of } x \end{cases}$$

is a possible probability density function. Find its cumulative distribution function.

The function $f(x)$ must satisfy conditions (a) and (c) above. This is the case since $f(x) \ge 0$ and

$$\int_{-\infty}^{\infty} f(x)dx = \int_a^b \frac{1}{b-a} dx = 1.$$

Also its cumulative distribution function $F(x)$ is given by

$$F(x) = \int_a^x \frac{1}{b-a} dx = \frac{x-a}{b-a} \quad \text{for} \quad a \le x \le b.$$

For $x < a$, $F(x) = 0$ and for $x > b$, $F(x) = 1$.

The pdf $f(x)$ is the density function of the **uniform distribution**.

1.6 Mean and variance

The **mean** (or **expectation** or **expected value**), $\mathbf{E}(X)$, of a discrete random variable X is defined as

$$\mu = \mathbf{E}(X) - \sum_{i=0}^{\infty} x_i p(x_i), \tag{1.7}$$

where $p(x_i) = \mathbf{P}(X = x_i)$, and, for a continuous random variable X, by

$$\mu = \mathbf{E}(X) = \int_{-\infty}^{\infty} x f(x)dx, \tag{1.8}$$

where $f(w)$ is the probability density function. It can be interpreted as the weighted average of the values of X in its sample space, where the weights are either the probability function or the density function. It is a measure which may be used to summarise the probability distribution of X in the sense that it is an average value. In the discrete case the summation over 'all x' includes both finite and infinite sample spaces.

A measure which is used in addition to the mean is the **variance** of X denoted by $\mathbf{V}(X)$ or σ^2. This gives a measure of variation or spread (dispersion) of the probability distribution of X, and is defined by

$$\begin{aligned} \sigma^2 &= \mathbf{V}(X) \\ &= \mathbf{E}[(X - \mathbf{E}(X))^2] = \mathbf{E}[(X - \mu)^2] \\ &= \begin{cases} \sum_{i=0}^{\infty} (x_i - \mu)^2 p(x_i) \text{ or } \sum_{x \in V} (x - \mu)^2 p(x), & \text{if } X \text{ is discrete,} \\ \int_{-\infty}^{\infty} (x - \mu)^2 f(x)dx, & \text{if } X \text{ is continuous.} \end{cases} \end{aligned} \tag{1.9}$$

The variance is the mean of the squared deviations of each value of X from the central value μ. In order to give a measure of variation which is in the same units as the mean, the square root σ of $\mathbf{V}(X)$ is used, namely

$$\sigma = \sqrt{\mathbf{V}(X)}. \tag{1.10}$$

This is known as the **standard deviation** (sd) of X.

A function of a random variable is itself a random variable. If $h(X)$ is a function of the random variable X, then it can be shown that the expectation of $h(X)$ is given by

$$\mathbf{E}[h(X)] = \begin{cases} \sum\limits_{i=0}^{\infty} h(x_i)p(x_i) \text{ or } \sum\limits_{x \in \mathcal{V}} h(x)p(x), & \text{if } X \text{ is discrete} \\ \int\limits_{-\infty}^{\infty} h(x)p(x)dx, & \text{if } X \text{ is continuous.} \end{cases}$$

It is relatively straightforward to derive the following results for the expectation and variance of a linear function of X:

$$\mathbf{E}(aX + b) = a\mathbf{E}(X) + b = a\mu + b, \tag{1.11}$$

$$\begin{aligned} \mathbf{V}(aX + b) &= \mathbf{E}[(aX + b - a\mu - b)^2] \\ &= \mathbf{E}[(aX - a\mu)^2] = a^2\mathbf{E}[(X - \mu)^2] = a^2\mathbf{V}(X), \quad (1.12) \end{aligned}$$

where a and b are constants. Note that the *translation* of the random variable does not affect the variance. Also

$$\mathbf{V}(X) = \mathbf{E}[(X - \mu)^2] = \mathbf{E}(X^2) - 2\mu\mathbf{E}(X) + \mu^2 = \mathbf{E}(X^2) - \mu^2, \tag{1.13}$$

which is sometimes known as the *computational formula for the variance*. These results hold whether X is discrete or continuous and enable us to interpret expectation and variance as operators on X. Of course, in all cases mean and variance only exist where the summations and infinite integrals are finite.

For expectations, it can be shown more generally that

$$\mathbf{E}\left[\sum_{i=1}^{k} a_i h_i(X)\right] = \sum_{i=1}^{k} a_i \mathbf{E}[h_i(X)], \tag{1.14}$$

where a_i, $i = 1, 2, \ldots, k$ are constants and $h_i(X)$, $i = 1, 2, \ldots, k$ are functions of the random variable X.

1.7 Some standard discrete probability distributions

In this section we shall look at discrete random variables X with probability mass function $p(x_i)$ or p_i, where x_i takes integer values. Each of the random variables considered are numerical outcomes of independent repetitions (or trials) of a simple experiment knowns as a **Bernoulli experiment**[3]. This is an experiment where there

---

[3] Jacob Bernoulli (1655–1705), Swiss mathematician.

are only two outcomes: a 'success' ($X = 1$) with probability p or a 'failure' ($X = 0$) with probability $q = 1 - p$. The value of the random variable X is used as an **indicator** of the outcome, which may also be interpreted as the presence or absence of a particular characteristic. For example, in a single coin toss $X = 1$ is associated with the occurrence, or the presence of the characteristic, of a *head*, and $X = 0$ with a *tail*, or the absence of a head.

The probability function of this random variable may be expressed as

$$p_1 = \mathbf{P}(X = 1) = p, \qquad p_0 = \mathbf{P}(X = 0) = q, \qquad (1.15)$$

where p is known as the **parameter** of the probability distribution, and $q = 1 - p$. We say that the random variable has a **Bernoulli distribution**.

The expected value of X is easily seen to be

$$\mu = \mathbf{E}(X) = 0 \times q + 1 \times p = p, \qquad (1.16)$$

and the variance of X is

$$\sigma^2 = \mathbf{V}(X) = \mathbf{E}(X^2) - \mu^2 = 0^2 \times q + 1^2 \times p - p^2 = pq. \qquad (1.17)$$

Suppose now that we are interested in random variables associated with independent repetitions of Bernoulli experiments, each with a probability of success, p. Consider first the probability distribution of a random variable X (a different X from that defined above) which is the number of successes in a fixed number of independent trials, n. If there are k successes and $n - k$ failures in n trials, then each sequence of 1's and 0's has the probability $\mathbf{P}(X = k) = p^k q^{n-k}$. The number of ways in which x successes can be arranged in n trials is the binomial expression

$$\frac{n!}{k!(n-k)!}, \quad \text{also expressed in the notation} \quad \binom{n}{k}.$$

Since each of these mutually exclusive sequences occurs with probability $p^k q^{n-k}$, the probability function of this random variable is given by

$$p_k = \binom{n}{k} p^k q^{n-k}, \qquad k = 0, 1, 2, \ldots, n. \qquad (1.18)$$

In (1.18) these are the $(n + 1)$ terms in the binomial expansion of $(p + q)^n$. For this reason, it is known as the **binomial distribution**, with parameters n and p. A consequence of this observation is that, as we expect,

$$\sum_{k=0}^{n} p_k = \sum_{k=1}^{n} \binom{n}{k} p^k q^{n-k} = (p + q)^n = 1.$$

The mean and variance may be easily shown to be np and npq, respectively, which is n times the mean and variance of the Bernoulli distribution (see Example 1.8).

Suppose that the random variable X is now the number k until the first success occurs:

$$p_k = \mathbf{P}(X = k) = q^{k-1} p, \quad k = 1, 2, \ldots,$$

that is, $(k - 1)$ failures before the first success. This is the **geometric distribution** with parameter p. This process is known as **inverse sampling**. Note that successive probabilities form a geometric series with common ratio $q = 1 - p$. Note that the sample space is now countably infinite. After some algebra it can be shown that the mean is given by $\mu = 1/p$, and the variance by $\sigma^2 = q/p^2$ (see Problem 1.9).

The geometric probability distribution possesses an interesting property known as the 'no memory', which can be expressed by

$$\mathbf{P}(X > a + b | X > a) = \mathbf{P}(X > b),$$

where a and b are positive integers. What this means is that if a particular event has *not* occurred in the first a repetitions of the experiment, then the probability that it will occur in the next b repetitions is the same as in the first b repetitions of the experiment. The result can be proved as follows, using the definition of conditional probability in Section 1.3:

$$\mathbf{P}(X > a + b | X > a) = \frac{\mathbf{P}(X > a + b \cap X > a)}{\mathbf{P}(X > a)} = \frac{\mathbf{P}(X > a + b)}{\mathbf{P}(X > a)}.$$

Since $\mathbf{P}(X > x) = q^x$,

$$\mathbf{P}(X > a + b | X > a) = \frac{q^{a+b}}{q^a} = q^b = \mathbf{P}(X > b).$$

The converse is also true, but the proof is not given here.

Consider now the case in which $r(> 1)$ successes have occurred. The probability function of the number of trials may be derived by noting that $X = k$ requires that the k-th trial results in the r-th success and that the remaining $r - 1$ successes may occur in any order in the previous $k - 1$ trials. The number of combinations in which $r - 1$ successes occur in $k - 1$ trials is the binomial

$$\frac{(k - 1)!}{(r - 1)!(k - r)!} = \binom{k - 1}{r - 1}.$$

Hence

$$p_k = \binom{k - 1}{r - 1} p^r q^{k-r}, \qquad k = r, r + 1, \ldots. \tag{1.19}$$

This is known as a **Pascal**[4] or **negative binomial distribution** with parameters r and p. Its mean is r/p and its variance rq/p^2, which are respectively r times the mean and r times the variance of the geometric distribution. Hence a similar relationship exists between the geometric and the Pascal distributions as between the Bernoulli and the binomial distributions.

The binomial random variable arises as the result of observing n independent identically distributed Bernoulli random variables, and the Pascal by observing r sets of geometric random variables.

Certain problems involve the counting of the number of events which have occurred in a fixed time period; for example, the number of emissions of alpha particles by an X-ray source or the number of arrivals of customers joining a queue. It has been found that the **Poisson distribution** is appropriate in modeling these counts when the underlying process generating them is considered to be completely random. We shall

---

[4] Blaise Pascal (1623–1662), French mathematician and scientist.

spend some time in Chapter 5 defining such a process, which is known as a **Poisson process**.

As well as being a probability distribution in its own right, the Poisson distribution also provides a convenient approximation to the binomial distribution to which it converges when n is large and p is small and $np = \alpha$, a constant. This is a situation where rounding errors would be likely to cause computational problems, if the numerical probabilities were to be calculated.

The Poisson probability function with parameter α is

$$p_k = \frac{e^{-\alpha}\alpha^k}{k!}, \qquad k = 0, 1, 2, \ldots$$

with mean and variance both equal to α.

The **discrete uniform distribution** with integer parameter n has a random variable X which can take the values $r, r+1, r+2, \ldots r+n-1$ with the same probability $1/n$ (the continuous uniform distribution was introduced in Example 1.6). Hence

$$\mathbf{P}(X - k) = \frac{1}{n} \qquad (k = r, r+1, \ldots r+n-1) \tag{1.20}$$

for any integer r. It is easy to show that the mean and variance of X are given by

$$\mu = r + \tfrac{1}{2}(n+1), \qquad \sigma^2 = \mathbf{V}(X) = \tfrac{1}{12}(n^2 - 1). \tag{1.21}$$

A simple example of the uniform discrete distribution is the fair die in which the faces $1, 2, 3, 4, 5, 6$ are equally likely to appear, each with probability $\tfrac{1}{6}$.

1.8 Some standard continuous probability distributions

The **exponential distribution** is a continuous distribution which will be used in subsequent chapters to model the random variable, say X, which is the *time* to a particular event. In a Poisson process (discussed in Chapter 5), we shall see that this is the time between successive occurrences of the event of interest. For example, the inter-arrival time of customers in a queue, or the lifetime or the time to failure of a component, where the failure rate α is assumed to be constant in reliability theory, can be modelled by exponential distributions.

The density function is

$$f(x) = \begin{cases} \alpha e^{-\alpha x}, & x \geq 0 \\ 0 & x < 0 \end{cases}$$

where α is a positive parameter. It is a density function since

$$\int_{-\infty}^{\infty} f(x)\mathrm{d}x = \int_0^{\infty} \alpha e^{-\alpha x}\mathrm{d}x = 1.$$

The mean and variance are given by

$$\mu = \mathbf{E}(X) = \int_0^{\infty} \alpha x e^{-\alpha x}\, dx = \frac{1}{\alpha}, \tag{1.22}$$

and

$$\sigma^2 = \mathbf{V}(X) = \int_0^\infty \left(x - \frac{1}{\alpha}\right)^2 \alpha e^{-\alpha x} dx = \frac{1}{\alpha^2}. \qquad (1.23)$$

The distribution function $F(x)$ of this random variable is given by

$$F(x) = \int_{-\infty}^x f(u)du = \int_0^x \alpha e^{-\alpha y} dy = 1 - e^{-\alpha x}.$$

If X has an exponential distribution, then using the conditional probability result from Section 1.3,

$$\mathbf{P}[X > a + b | X > a] = \frac{\mathbf{P}[X > a + b]}{\mathbf{P}[X > a]} = \frac{e^{-\alpha(a+b)}}{e^{-\alpha a}} = e^{\alpha b}, \qquad (1.24)$$

where $a, b > 0$. This shows the 'no-memory' property of the exponential distribution, that is, the result does not depend on a. It can be shown that the exponential distribution is the only *continuous* distribution with this property.

A random variable X has a **normal distribution** with two parameters, the mean μ and variance $\mathbf{V}(X)$ or σ^2, if its probability density function is

$$f(x) = \frac{1}{\sigma\sqrt{2\pi}} \exp\left[-\frac{(x - \mu)^2}{2\sigma^2}\right] \qquad (-\infty < x < \infty). \qquad (1.25)$$

That

$$\int_{-\infty}^\infty f(x)dx = 1$$

follows from the standard integral

$$\int_{-\infty}^\infty e^{-u^2} du = \sqrt{\pi}.$$

For the normal distribution it can be confirmed that the mean and variance are

$$\int_{-\infty}^\infty x f(x)dx = \mathbf{E}(X) = \mu, \qquad \int_{-\infty}^\infty (x - \mu)^2 f(x)dx = \sigma^2 = \mathbf{V}(X). \qquad (1.26)$$

The normal distribution is denoted by $N(\mu, \sigma^2)$.

By using a change of variable, then $Z = (X - \mu)/\sigma$ is easily shown to be normally distributed with mean 0 and variance 1, or $N(0, 1)$ with density (see Section 1.6 and Problem 1.28(b) for an alternative derivation)

$$f(z) = \frac{1}{\sqrt{2\pi}} \exp[-\tfrac{1}{2}z^2], \quad -\infty < z < \infty. \qquad (1.27)$$

The transformation of X is known as **standardising**, and consequently Z is termed a **standard normal variable**. The graph of $N(0, 1)$ is shown in Figure 1.5.[5] This transformation enables calculation of probabilities for *any* normally distributed random variable. Suppose, for example, X is a $N(2, 9)$ random variable, then

$$\mathbf{P}(-1 < X < 3) = \mathbf{P}(-1 < Z < \tfrac{1}{3}) = \Phi(\tfrac{1}{3}) - \Phi(-1) = 0.629 - 0.159 = 0.470$$

(see Figure 1.5).

---

[5] Tables of the cumulative function $\mathbf{P}(Z < z) = \Phi(z)$ can be found on the internet, or in most statistics texts.

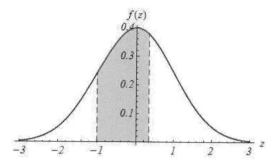

Figure 1.5 *Standardised normal distribution* $N[0, 1]$.

Consider now the random variable $Y = aX + b$, where a and b are constants: Y is a linear function of X. Again using a change of variable it may be shown that Y is normally distributed with mean $a\mu + b$ and variance $a^2\sigma^2$, or $N(a\mu + b, a^2\sigma^2)$ (see Problem 1.28(c) for an alternative derivation). The distribution Z may be recovered by setting $a = 1/\mu$ and $b = -\mu/\sigma$.

The **gamma distribution** depends on the properties of the gamma function $\Gamma(n)$ defined by

$$\Gamma(n) = \int_0^\infty x^{n-1}e^{-x}dx. \tag{1.28}$$

Integration by parts produces the recurrence formula

$$\Gamma(n) = (n-1)\Gamma(n-1),$$

since

$$\Gamma(n) \quad - \quad -\int_0^\infty x^{n-1}\frac{de^{-x}}{dx}dx$$

$$= \quad [-x^{n-1}e^{-x}]_0^\infty + \int_0^\infty \frac{dx^{n-1}}{dx}e^{-x}dx$$

$$= \quad (n-1)\int_0^\infty x^{n-2}e^{-x}dx = (n-1)\Gamma(n-1).$$

When n is an integer it follows that

$$\Gamma(n) = (n-1)!.$$

The gamma distribution has two parameters $n > 0$, $\alpha > 0$, and has the density function

$$f(x) = \frac{\alpha^n}{\Gamma(n)}x^{n-1}e^{-\alpha x}, \qquad x > 0.$$

Note that by setting $n = 1$, we obtain the exponential distribution. The mean is given by

$$\mathbf{E}(X) = \frac{\alpha^n}{\Gamma(n)}\int_0^\infty x^n e^{-\alpha x}dx = \frac{\alpha^n}{\Gamma(n)\alpha^{n+1}}\int_0^\infty y^n e^{-y}dy,$$

after the change of variable $y = \alpha x$. Hence

$$\mathbf{E}(X) = \frac{\Gamma(n+1)}{\alpha \Gamma(n)} = \frac{n}{\alpha}.$$

It can be shown that the variance of the gamma distribution is n/α^2. Note that the mean and variance are respectively n times the mean and variance of the exponential distribution with parameter α.

It will be shown in the next section that if $X_1, X_2, \ldots, X_n$ are independent identically distributed exponential random variables, then

$$Y = \sum_{i=1}^{n} X_i$$

has a gamma distribution with parameters α, n.

Another distribution arising in reliability (Chapter 8) is the **Weibull distribution**[6] which depends on two positive parameters, α and β, and has density

$$f(x) = \alpha \beta x^{\beta-1} e^{-\alpha x^\beta}, \qquad x > 0.$$

This enables more complex lifetime data to be modeled, especially where a constant failure rate is not a reasonable assumption. Note that setting $\beta = 1$ gives the exponential distribution. After some algebra, it may be shown that the mean and variance of the Weibull distribution are

$$\mu = \frac{1}{\alpha^{\frac{1}{\beta}}} \Gamma\left(\frac{1}{\beta} + 1\right),$$

$$\sigma^2 = \frac{1}{\alpha^{\frac{2}{\beta}}} \left[\Gamma\left(\frac{2}{\beta} + 1\right) - \left\{\Gamma\left(\frac{1}{\beta} + 1\right)\right\}^2\right].$$

1.9 Generating functions

In this section we are going to consider two generating functions. The main purpose is to use their uniqueness properties to identify the probability distributions of functions of random variables. The first is the **moment generating function** (mgf). This function depends on a dummy variable s and uses the series expansion of e^{sX} to generate the **moments** $\mathbf{E}(X^r)$, $(r \geq 1)$ of the random variable X. The expected value $\mathbf{E}(X^r)$ is called the r-th moment of the random variable X about zero. The power series expansion of e^{sX} is given by

$$e^{sX} = 1 + sX + \frac{(sX)^2}{2!} + \cdots + \frac{(sX)^r}{r!} + \cdots,$$

a series which converges for all sX. The mgf is obtained by taking expected values of both sides of this equation,

$$M_X(s) = \mathbf{E}(e^{sX}) = \mathbf{E}\left[1 + sX + \frac{(sX)^2}{2!} + \cdots + \frac{(sX)^r}{r!} + \cdots\right]. \qquad (1.29)$$

---

[6] Waloddi Weibull (1887–1979), Swedish engineer.

We mentioned in Section 1.6 that the expected value of a finite sum of functions was equal to the sum of the expected values, and that $\mathbf{E}(aX) = a\mathbf{E}(X)$. We now wish to apply this result to an infinite sum, making the assumption that this result holds (it does so under fairly general conditions). Thus we assume that

$$M_X(s) = 1 + s\mathbf{E}(X) + \frac{s^2}{2!}\mathbf{E}(X^2) + \cdots + \frac{s^r}{r!}\mathbf{E}(X^r) + \cdots = \sum_{r=0}^{\infty} \frac{s^r}{r!}\mathbf{E}(X^r). \quad (1.30)$$

The coefficient of $s^r/r!$ is therefore the r-th moment of X. Taking successive derivatives of the mgf with respect to s, and then setting $s = 0$, we can obtain these moments. For example,

$$M_X'(0) = \mathbf{E}(X) = \boldsymbol{\mu}, \qquad M_X''(0) = \mathbf{E}(X^2),$$

and the variance is therefore given by (see Section 1.6)

$$\sigma^2 = M_X''(0) - [M_X'(0)]^2.$$

Let X have a gamma distribution with parameters n, α. Then, using the substitution $x = u/(\alpha - s)$,

$$\begin{aligned} M_X(s) = \mathbf{E}(e^{sX}) &= \int_0^{\infty} \frac{\alpha^n}{\Gamma(n)} e^{sx} x^{n-1} e^{-\alpha x} dx, \\ &= \int_0^{\infty} \frac{\alpha^n}{\Gamma(n)} x^{n-1} e^{-x(\alpha - s)} dx. \\ &= \frac{\alpha^n}{\Gamma(n)(\alpha - s)^n} \int_0^{\infty} u^{n-1} e^{-u} du = \left(\frac{\alpha}{\alpha - s}\right)^n \end{aligned}$$

provided that $s < \alpha$. Now consider the result quoted but not proved in the previous section on the distribution of **independent and identically distributed (iid)** exponential random variables $X_1, X_2, \ldots, X_n$.

We may now use the two results above to prove this. We first need to note that since the random variables are independent, it follows that

$$\mathbf{E}[g_1(X_1)g_2(X_2) \cdots g_n(X_n)] = \mathbf{E}[g_1(X_1)]\mathbf{E}[g_2(X_2)] \cdots \mathbf{E}[g_n(X_n)]$$

for random variable functions $g_1(X_1), g_2(X_2), \ldots g_n(X_n)$. Let $Y = \sum_{i=1}^n X_i$. Then

$$\begin{aligned} M_Y(s) &= \mathbf{E}[\exp(sY)] = \mathbf{E}\left[\exp\left(s\sum_{i=1}^n X_i\right)\right] \\ &= \mathbf{E}[\prod_{i=1}^n \exp(sX_i)] = \prod_{i=1}^n \mathbf{E}[\exp(sX_i)] \quad \text{(using the result above)} \\ &= \prod_{i=1}^n M_{X_i}(s), \quad\quad\quad\quad\quad\quad\quad\quad\quad\quad\quad\quad\quad\quad\quad (1.31) \end{aligned}$$

but since the X_is are identically distributed, they have the same mgf $M_X(s)$. Hence

$$M_Y(s) = \left[M_X(s)\right]^n.$$

Of course this result holds for any iid random variables. If X is exponential, then

$$M_X(s) = \frac{\alpha}{\alpha - s},$$

and

$$M_Y(s) = \left(\frac{\alpha}{\alpha - s}\right)^n,$$

which is the mgf of a gamma-distributed random variable.

Example 1.7 *n rails are cut to a nominal length of ℓ meters. By design the cutting machine cannot cut lengths less than ℓ but the actual length of any rail can have positive error of x_i, which is exponentially distributed with parameter α independently of each other. The rails are welded end-to-end to form a continuous rail of nominal length $n\ell$. Find the expected value and variance of the length of the composite rail.*

Let X_i be the random variable representing the error x_i of rail i. Let Y be the random variable of the additional length of the whole rail so that

$$Y = X_1 + X_2 + \cdots + X_n.$$

Since the random variables are independent and identically distributed, and the errors exponentially distributed, the moment generating function is (see above)

$$M_Y(s) = \prod_{i=1}^{n} M_{X_i}(s) = [M_{X_1}(s)]^n.$$

Therefore

$$
\begin{aligned}
M_Y(s) &= \left(\frac{\alpha}{\alpha - s}\right)^n = \left(1 - \frac{s}{\alpha}\right)^{-n} \\
&= 1 + \frac{n}{\alpha}s + \frac{1}{2}\frac{n(n+1)}{\alpha^2}s^2 + \cdots.
\end{aligned}
$$

Hence the expected value of the length of the composite rail and the variance of the error are given by

$$\mathbf{E}(Y) = \frac{n}{\alpha}, \qquad \sigma^2 = \mathbf{E}(Y^2) - [\mathbf{E}(Y)]^2 = \frac{n(n+1)}{\alpha^2} - \frac{n^2}{\alpha^2} = \frac{n}{\alpha^2}.$$

The latter result is an example of the result that

$$\mathbf{V}(X_1 + X_2 + \cdots + X_n) = \mathbf{V}(X_1) + \mathbf{V}(X_2) + \cdots + \mathbf{V}(X_n),$$

provided $X_1, X_2, \ldots, X_n$ are independent.

The moment generating function (mgf) M_X of a $N(\mu, \sigma^2)$ variable is

$$\exp(\mu s + \tfrac{1}{2}\sigma^2 s^2) \tag{1.32}$$

(see Problem 1.28(a)). Hence by inspection the mgf of a standard normal Z is

$$M_Z(s) = \exp(\tfrac{1}{2}s^2). \tag{1.33}$$

Consider a sequence of n iid $N(\mu, \sigma^2)$ random variables $X_1, X_2, \ldots, X_n$. Then for the random variable $Y = \sum_{i=1}^{n} X_i$, using the results above,

$$M_Y(s) = [\exp(\mu s + \tfrac{1}{2}\sigma^2 s^2)]^n = \exp(n\mu s + \tfrac{1}{2}n\sigma^2 s^2), \tag{1.34}$$

which is the mgf of an $N(n\mu, n\sigma^2)$ random variable. Hence by the uniqueness property of mgf's, Y is normally distributed with mean $n\mu$ and variance $n\sigma^2$.

If the sequence of independent normally distributed random variables now have different means and, say, variances μ_1 and σ_i^2, then Y will be normally distributed with mean $\sum_{i=1}^{n} \mu_i$ and variance $\sum_{i=1}^{n} \sigma_i^n$ (see Problem 1.28(d)). A further generalisation is to consider a linear sum of these random variables $W = \sum_{i=1}^{n} a_i X_i$ where the a_i's are constants. It may be shown that W is normally distributed with mean $\sum_{i=1}^{n} a_i \mu_i$ and variance $\sum_{i=1}^{n} a_i^2 \sigma_i^2$ (see Problem 1.28(e)).

Moment generating functions may be defined for both discrete and continuous random variables, but the **probability generating function** (pgf) is only defined for integer-valued random variables. To be specific, we consider the probability distribution $p_n = \mathbf{P}(N = n)$, where the random variable N can only take the values $0, 1, 2, \ldots$. (We use the notation p_n rather than $p(n)$ in this context since it conforms to the usual notation for coefficients in power series.) Again it is expressed in terms of a power series in a dummy variable s, and is defined as the expected value of s^N:

$$G_N(s) = \mathbf{E}(s^N) = \sum_{n=0}^{\infty} p_n s^n, \tag{1.35}$$

provided the right-hand side converges. The question of uniqueness arises with generating functions, since we deduce distributions from them. Later we shall represent the probability generating function by $G(s)$ without the random variable subscript. It can be shown (see Grimmett and Welsh (1986)) that two random variables X and Y have the same probability generating function if and only if they have the same probability distributions.

In many cases the pgf will be a polynomial: this will occur if the outcomes can only be a finite as in a death process. In others, such as in birth and death processes (Chapter 6), in which there is, theoretically, no upper bound to the population size, the series for the pgf will be infinite. If a pgf is an infinite power series there are requirements for the convergence, uniqueness, and term-by-term differentiation of the series, but we shall assume here that such conditions are generally met without further discussion.

With these assumption, if $G_N(s) = \sum_{n=0}^{\infty} p_n s^n$, then

$$\frac{dG_N}{ds} = G'_N(s) = \sum_{n=1}^{\infty} n p_n s^n, \quad (0 \le s \le 1),$$

$$\frac{d^2 G_N}{ds^2} = G''_N(s) = \sum_{n=2}^{\infty} n(n-1) p_n s^{n-2}, \quad (0 \le s \le 1).$$

The pgf has the following properties:

- (a) $G_N(1) = \sum_{n=0}^{\infty} p_n = 1$.

- (b) $G'_N(1) = \sum_{n=0}^{\infty} n p_n = \mathbf{E}(N) = \mu$, the mean.

- (c) $G_N''(1) = \sum_{n=0}^{\infty} n(n-1)p_n = \mathbf{E}(N^2) - \mathbf{E}(N)$, so that the variance is given by

$$\mathbf{V}(N) = \sigma^2 = \mathbf{E}(N^2) - \mu^2 = G''(1) + G'(1) - [G'(1)]^2.$$

- (d) $G_N^{(m)}(1) = d^m G(1)/ds^m = \mathbf{E}[N(N-1)\ldots(N-m+1)]$, which is called the **factorial moment** of N.

- (e) If $N_1, N_2, \ldots, N_r$ are independent and identically distributed discrete random variables, and $Y = \sum_{i=1}^{r} N_i$, then

$$
\begin{aligned}
G_Y(s) &= \mathbf{E}[s^Y] = \mathbf{E}\left[s^{\sum_{i=1}^{r} N_i}\right] = \mathbf{E}\left[\prod_{i=1}^{r}(s^{N_i})\right] \\
&= \prod_{i=1}^{r} G_{N_i}(s),
\end{aligned}
$$

and since the N_i's are identically distributed,

$$G_Y(s) = [G_X(s)]^r, \qquad (1.36)$$

which is similar to the result for moment generating functions.

Example 1.8 *The random variable N has a binomial distribution with parameters m, p. Its probability function is given by*

$$p(n) = p_n = \mathbf{P}(N = n) = \binom{m}{n} p^n q^{m-n}, \qquad n = 0, 1, 2, \ldots, m$$

(see Eqn (1.23)). Find its pgf, and its mean and variance.

The pgf of N is

$$
\begin{aligned}
G_N(s) = G(s) &= \sum_{n=0}^{m} s^n \binom{m}{n} p^n q^{m-n} = \sum_{n=0}^{m} \binom{m}{n} (ps)^n q^{m-n} \\
&= (q + ps)^m
\end{aligned}
$$

using the binomial theorem. It follows that

$$G'(s) = mp(q + ps)^{m-1},$$
$$G''(s) = m(m-1)p^2(q + ps)^{m-2}.$$

Using the results above, the mean and variance are given by

$$\mu = G'(1) = mp,$$

and

$$\sigma^2 = G''(1) + G'(1) - [G'(1)]^2 = m(m-1)p^2 + mp - m^2 p^2 = mpq.$$

The Bernoulli distribution is the binomial distribution with $m = 1$. Hence its pgf is

$$G_X(s) = q + ps.$$

Consider n independent and identically distributed Bernoulli random variables $X_1, X_2, \ldots, X_n$, and let $Y = \sum_{i=1}^{n} X_i$. Then from the results above,

$$G_Y(s) = [G_X(s)]^n = (q + ps)^n,$$

which is again the pgf of a binomial random variable.

It is possible to associate a generating function with any sequence $\{a_n\}$, $(n = 0, 1, 2, \ldots)$ in the form

$$H(s) = \sum_{n=0}^{\infty} a_n s^n$$

provided that the series converges in some interval containing the origin $s = 0$. Unlike the pgf, this series need not satisfy the conditions $H(1) = 1$ for a probability distribution nor $0 \leq a_n \leq 1$. An application using such a series is given in Problem 1.24, and in Section 3.3 on random walks.

1.10 Conditional expectation

In many applications of probability we are interested in the possible values of two or more characteristics in a problem. For this we require two or more random variables which may or may not be independent. We shall only consider the case of two discrete random variables X and Y which form a two-dimensional random variable denoted by (X, Y), which can take pairs of values (x_i, y_j): for example, we could assume $(i = 1, 2, \ldots; j = 1, 2, \ldots)$ (either sequence may be finite or infinite, or start for some other values) with joint probability (mass) function $p(x_i, y_j)$, which is now a function of two variables. As in Section 1.4, the probabilities must satisfy

- (a) $0 \leq p(x_i, y_j) \leq 1$ for all (i, j),

- (b) $\displaystyle\sum_{i \varepsilon I} \sum_{j \varepsilon J} p(x_i, y_j) = 1$.

The domains of i and j are defined by $i \varepsilon I$ and $j \varepsilon J$ where I and J can be either bounded or unbounded sequences of the integers.

The random variables X and Y are said to be **independent**, if and only if,

$$p(x_i, y_j) = q(x_i) r(y_j) \quad \text{for all } i \text{ and } j, \tag{1.37}$$

where, of course,

$$\sum_{i \varepsilon I} q(x_i) = 1, \quad \sum_{j \varepsilon J} r(y_j) = 1.$$

If the random variable $Z = H(X, Y)$ is a function of the random variables X and Y, then we will state without proof that the expected value of Z is given by

$$\mathbf{E}(Z) = \sum_{i \varepsilon I} \sum_{j \varepsilon J} H(x_i, y_j) p(x_i, y_j).$$

We now introduce the concept of **conditional expectation**. This is an informal introduction with the intention of presenting the main ideas in a comprehensible manner. This comes with a caution. There are aspects such as probability spaces,

measure theory, existence, and convergence, which will be assumed here but will not
be explicitly defined.

We consider the joint distribution of two random variables X and Y with joint
mass function $p(x_i, y_j)$. We assume that we can associate with X and Y conditional
probabilities

$$\mathbf{P}(X = x_i | Y = y_j) \text{ and } \mathbf{P}(Y = y_j | x = x_i).$$

We use alternative equivalent notations

$$p_X(x_i | y_j) \text{ or simply } \mathbf{P}(X|Y)$$

if the context is clear for the first, and similarly for the second, namely

$$p_Y(y_j | x_i) \text{ or simply } \mathbf{P}(Y|X)$$

(there are a plethora of notations in this subject which make comparisons between
texts hard work). With these probabilities we can consider expectations. Hence the
conditional expectation of X for each element of Y is

$$\mathbf{E}(X|Y = y_j) = \sum_{i \in I} x_i \mathbf{P}(X = x_i | Y = y_j) \text{ or } \sum_{i \in I} x_i p_X(x_i | y_j). \qquad (1.38)$$

The significance of this expectation is that it has value *for each element* of J. The
totality of these values is $\mathbf{E}[X|Y]$, which is a *random variable*. It will have the same
number of elements as J. Similarly

$$\mathbf{E}(Y|X = x_i) = \sum_{j \in J} y_j p_Y(y_j | x_i), \qquad (1.39)$$

and $\mathbf{E}[Y|X]$ is another random variable.

The conditional probabilities are given by

$$p_X(x_i | y_j) = p(x_i, y_j) / \sum_{i \in I} p(x_i, y_j), \qquad (1.40)$$

and

$$p_Y(y_j | x_i) = p(x_i, y_j) / \sum_{j \in J} p(x_i, y_j). \qquad (1.41)$$

In (1.40), $p(x_i, y_j)$ is the probability that X and Y occur (intersection in Section
1.3), and the denominator is the probability that Y occurs. A similar interpretation
applies to (1.41). The probabilities $\sum_{i \in I} p(x_i, y_j)$ and $\sum_{j \in J} p(x_i, y_j)$ are known as
the **marginal probability distributions**.

If X and Y are independent then (1.37) holds, and (1.40) and (1.41) become simply

$$p_X(x_i, y_j) = q(x_i), \qquad p_Y(x_i, y_j) = r(y_j),$$

with expectations

$$\mathbf{E}(X|Y) = \sum_{i \in I} x_i q(x_i) = \mathbf{E}(X),$$

$$\mathbf{E}(Y|X) = \sum_{j \in J} y_j r(y_j) = \mathbf{E}(Y).$$

Returning to the dependent case, we can anticipate that the conditional expectations $\mathbf{E}(X|Y)$ and $\mathbf{E}(Y|X)$, being random variables, will also have expectations. Thus

$$
\begin{aligned}
\mathbf{E}(\mathbf{E}(X|Y)) &= \sum_{j \varepsilon J} \mathbf{E}(X|Y) \sum_{k \varepsilon I} p(x_k, y_j) \\
&= \sum_{j \varepsilon J} \sum_{i \varepsilon I} x_i p_X(x_i, y_j) \sum_{k \varepsilon I} p(x_k, y_j) \\
&= \sum_{j \varepsilon J} \sum_{i \varepsilon I} x_i p(x_i, y_j) \\
&= \mathbf{E}(X)
\end{aligned}
\tag{1.42}
$$

using (1.40) for the conditional probability. In words this states that the expected value of the conditional expectation of X with respect to Y is simply the expected value of X. Similarly,

$$
\mathbf{E}(\mathbf{E}(Y|X)) = \mathbf{E}(Y).
\tag{1.43}
$$

The following routine example, whilst of no intrinsic value, attempts in a particular case to identify the various probabilities and expectations, which are not always easily understood in the general theory.

Example 1.9 Let the random variable (X, Y) take the values (x_i, y_j) $(i = 1, 2, 3;\ j = 1, 2, 3)$. The mass functions $p(x_i, y_j)$ are given in the table.

$p(x_i, y_j)$	y_1	y_2	y_3
x_1	0.25	0	0.05
x_2	0.05	0.10	0.15
x_3	0.05	0.25	0.10

Find the random variable $\mathbf{E}(X|Y)$ and verify that

$$
\mathbf{E}(\mathbf{E}(X|Y)) - \mathbf{E}(X)
$$

We have gone into the calculations perhaps in a long-winded manner but it helps to understand the multiple summations which occur for conditional expectations. It could be a useful checklist. In this example $I = \{1, 2, 3\}$ and $J = \{1, 2, 3\}$. In the table, for example, $p(x_1, y_3) = 0.05$ and $p(x_3, y_2) = 0.25$. Note also that

$$
\sum_{i=1}^{3} \sum_{j=1}^{3} p(x_i, y_j) = 1
$$

as required.

The marginal probability distributions for Y will each have three components, namely

$$
\sum_{i=1}^{3} p(x_i, y_1) = p(x_1, y_1) + p(x_2, y_1) + p(x_3, y_1) = 0.25 + 0.05 + 0.05 = 0.35,
$$

$$
\sum_{i=1}^{3} p(x_i, y_2) = 0 + 0.1 + 0.25 = 0.35,
$$

$$\sum_{i=1}^{3} p(x_i, y_3) = 0.05 + 0.15 + 0.10 = 0.3.$$

Similarly for X,

$$\sum_{j=1}^{3} p(x_1, y_j) = 0.3, \quad \sum_{j=1}^{3} p(x_2, y_j) = 0.3, \quad \sum_{j=1}^{3} p(x_3, y_j) = 0.4.$$

Using (1.40) and (1.41) we can compute the conditional probabilities, which also require the marginal distributions previously calculated. Hence

$$p_X(x_1|y_1) = p(x_1, y_1) / \sum_{i=1}^{3} p(x_i, y_1) = 0.25/0.35 = 5/7,$$

$$p_X(x_2|y_1) = 0.05/0.35 = 1/7, \quad p_X(x_3|y_1) = 0.05/0.35 = 1/7.$$

$$p_X(x_1|y_2) = 0, \quad p_X(x_2|y_2) = 2/7, \quad p_X(x_3|y_2) = 5/7,$$

$$p_X(x_1|y_3) = 1/6, \quad p_X(x_2|y_3) = 1/2, \quad p_X(x_3|y_3) = 1/3.$$

From (1.38) we can find the component of the random variable, namely,

$$\mathbf{E}(X|Y = y_1) = \sum_{i=1}^{3} x_i p(x_i|y_1) = 1 \times \frac{5}{7} + 2 \times \frac{1}{7} + 3 \times \frac{1}{7} = \frac{10}{7},$$

$$\mathbf{E}(X|Y = y_2) = \sum_{i=1}^{3} x_i p(x_i|y_2) = \frac{19}{7},$$

$$\mathbf{E}(X|Y = y_3) = \sum_{i=1}^{3} x_i p(x_i|y_3) = \frac{13}{6},$$

Finally the random variable $\mathbf{E}(X|Y)$ takes the values

$$\left\{ \frac{10}{7}, \frac{19}{7}, \frac{13}{6} \right\} = Z = \{z_1, z_2, z_3\},$$

say.

The expected value of $\mathbf{E}(X|Y)$ or Z is

$$\mathbf{E}(Z) = \sum_{j=1}^{3} z_j \sum_{i=1}^{3} p(x_i, y_j) = \frac{10}{7} \times 0.35 + \frac{19}{7} \times 0.35 + \frac{13}{6} \times 0.3 = \frac{21}{10}.$$

Equally

$$\mathbf{E}(X) = \sum_{i=1}^{3} \sum_{j=1}^{3} x_i p(x_i, y_j) = 1 \times 0.3 + 2 \times 0.3 + 3 \times 0.4 = \frac{21}{10},$$

which confirms that $\mathbf{E}(\mathbf{E}(X|Y)) = \mathbf{E}(X)$.

These calculations can be repeated for $\mathbf{E}(Y|X)$ but this is left as a useful exercise in Problem 1.27.

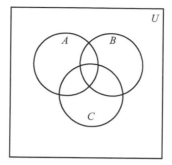

Figure 1.6 *See Problem 1.1.*

1.11 Problems

1.1. The Venn diagram of three events is shown in Figure 1.6. Indicate on the diagram
(a) $A \cup B$; (b) $A \cup (B \cup C)$; (c) $A \cap (B \cup C)$; (d) $(A \cap C)^c$; (e) $(A \cap B) \cup C^c$.

1.2. In a random experiment, A, B, C are three events. In set notation, write down expressions for the events:
(a) only A occurs;
(b) all three events A, B, C occur;
(c) A and B occur but C does not;
(d) at least one of the events A, B, C occurs;
(e) exactly one of the events A, B, C occurs;
(f) not more than two of the events occur.

1.3. For two events A and B, $\mathbf{P}(A) = 0.4$, $\mathbf{P}(B) - 0.5$, and $\mathbf{P}(A \cap B) - 0.3$. Calculate
(a) $\mathbf{P}(A \cup B)$; (b) $\mathbf{P}(A \cap B^c)$; (c) $\mathbf{P}(A^c \cup B^c)$.

1.4. Two distinguishable fair dice a and b are rolled. What are the elements of the sample space? What is the probability that the sum of the face values of the two dice is 9? What is the probability that at least one 5 or at least one 3 appears?

1.5. Two distinguishable fair dice are rolled. What is the probability that the sum of the faces is not more than 6?

1.6. For the probability generating function

$$G(s) = (2 - s)^{-\frac{1}{2}}$$

find the probability function $\{p_n\}$ and its mean.

1.7. Find the probability generating function $G(s)$ of the Poisson distribution (see Section 1.7) with parameter α given by

$$p_n = \frac{e^{-\alpha}\alpha^n}{n!}, \qquad n = 0, 1, 2, \ldots.$$

Determine the mean and variance of $\{p_n\}$ from the generating function.

1.8. A panel contains n warning lights. The times to failure of the lights are the independent random variables $T_1, T_2, \ldots, T_n$, which have exponential distributions with parameters $\alpha_1, \alpha_2, \ldots, \alpha_n$, respectively. Let T be the random variable of the time to first failure, that is,

$$T_i = \min\{T_1, T_2, \ldots, T_n\}.$$

Show that T_i has an exponential distribution with parameter $\sum_{j=1}^n \alpha_j$.

1.9. The geometric distribution with parameter p is given by

$$p(x) = q^{x-1}p, \quad x = 1, 2, \ldots$$

where $q = 1 - p$ (see Section 1.7). Find its probability generating function. Calculate the mean and variance of the geometric distribution from its pgf.

1.10. Two distinguishable fair dice a and b are rolled. What are the probabilities that:
(a) at least one 4 appears;
(b) only one 4 appears;
(c) the sum of the face values is 6;
(d) the sum of the face values is 5 and one 3 is shown;
(e) the sum of the face values is 5 or only one 3 is shown?

1.11. Two distinguishable fair dice a and b are rolled. What is the expected sum of the face values? What is the variance of the sum of the face values?

1.12. Three distinguishable fair dice a, b, and c are rolled. How many possible outcomes are there for the faces shown? When the dice are rolled, what is the probability that just two dice show the same face values and the third one is different?

1.13. In a sample space S, the events B and C are mutually exclusive, but A and B are not. Show that

$$\mathbf{P}(A \cup (B \cup C)) = \mathbf{P}(A) + \mathbf{P}(B) + \mathbf{P}(C) - \mathbf{P}(A \cap (B \cup C)).$$

From a well-shuffled pack of 52 playing cards a single card is randomly drawn. Find the probability that it is a club or an ace or the king of hearts.

1.14. Show that

$$f(x) = \begin{cases} 0 & x < 0 \\ 1/(2a) & 0 \leq x \leq a \\ e^{-(x-a)/a}/(2a) & x > a \end{cases}$$

is a possible probability density function. Find the corresponding cumulative distribution function.

1.15. A biased coin is tossed. The probability of a head is p. The coin is tossed until the first head appears. Let the random variable N be the total number of tosses including the first head. Find $\mathbf{P}(N = n)$, and its pgf $G(s)$. Find the expected value of the number of tosses.

1.16. The m random variables $X_1, X_2, \ldots, X_m$ are independent and identically distributed each with a gamma distribution with parameters n and α. The random variable Y is defined by

$$Y = X_1 + X_2 + \cdots + X_m.$$

Using the moment generating function, find the mean and variance of Y.

1.17. A probability generating function with parameter $0 < \alpha < 1$ is given by

$$G(s) = \frac{1 - \alpha(1 - s)}{1 + \alpha(1 - s)}.$$

Find $p_n = \mathbf{P}(N = n)$ by expanding the series in powers of s. What is the mean of the probability function $\{p_n\}$?

1.18. Find the moment generating function of the random variables X which has the uniform distribution

$$f(x) = \begin{cases} 1/(b - a), & a \le x \le b, \\ 0, & \text{for all other values of } x. \end{cases}$$

Deduce $\mathbf{E}(X^n)$.

1.19. A random variable X has a normal distribution with mean μ and variance σ^2. Find its moment generating function.

1.20. Find the probability generating functions of the following distributions, in which $0 < p < 1$:
(a) Bernoulli distribution: $p_n = p^n(1 - p)^{1-n}$, $(n = 0, 1)$;
(b) geometric distribution: $p_n = p(1 - p)^{n-1}$, $(n = 1, 2, \ldots)$;
(c) negative binomial distribution with parameters r and p expressed in the form:

$$p_n = \binom{r + n - 1}{r - 1} p^r (1 - p)^n, \qquad (n = 0, 1, 2, \ldots)$$

where r is a positive integer. In each case also find the mean and variance of the distribution using the probability generating function.

1.21. A word of five letters is transmitted by code to a receiver. The transmission signal is weak, and there is a 5% probability that any letter is in error independently of the others. What is the probability that the word is received correctly? The same word is transmitted a second time with the same errors in the signal. If the same result is received, what is the probability now that the word is correct?

1.22. A binary code of 500 bits is transmitted across a weak link. The probability that any bit has a transmission error is 0.0004 independently of the others.
  (a) What is the probability that only the first bit fails?
  (b) What is the probability that the code is transmitted successfully?
  (c) What is the probability that at least two bits fail?

1.23. The source of a beam of light is a perpendicular distance d from a wall of length $2a$, with the perpendicular from the source meeting the wall at its midpoint. The source emits a pulse of light randomly in a direction θ, the angle between the direction of the pulse and the perpendicular, chosen uniformly in the range $-\tan^{-1}(a/d) \le \theta \le \tan^{-1}(a/d)$. Find the probability distribution of x $(-a \le x \le a)$, where the pulses hit the wall. Show that its density function is given by

$$f(x) = \frac{d}{2(x^2 + d^2)\tan^{-1}(a/d)}$$

(this the density function of a **Cauchy distribution**[7]). If $a \to \infty$, what can you say about the mean of this distribution?

1.24. Suppose that the random variable X can take the integer values $0, 1, 2, \ldots$. Let p_j and q_j be the probabilities

$$p_j = \mathbf{P}(X = j), \quad q_j = \mathbf{P}(X > j), \quad (j = 0, 1, 2, \ldots).$$

Show that, if

$$G(s) = \sum_{j=0}^{\infty} p_j s^j, \qquad H(s) = \sum_{j=0}^{\infty} q_j s^j,$$

then $(1 - s)H(s) = 1 - G(s)$.
   Show also that $\mathbf{E}(X) = H(1)$.

1.25 In a lottery, players can choose q numbers from the consecutive integers $1, 2, \ldots, n$ ($q < n$). The player wins if r numbers ($3 \leq r \leq q$) agree with the r numbers randomly chosen from the n integers. Show that the probability of r numbers being correct is

$$\binom{q}{r}\binom{n-q}{q-r} \Big/ \binom{n}{q}.$$

Compute the probabilities if $n = 49$, $q = 6$, $r = 3, 4, 5, 6$ (the UK lottery).

1.26. A count of the second edition of this book showed that it contains 181,142 Roman letters (not case sensitive: Greek not included). The table of the frequency of the letters:

Frequency table

a	13011	j	561	s	12327
b	4687	k	1424	t	15074
c	6499	l	6916	u	5708
d	5943	m	5487	v	2742
e	15273	n	14265	w	3370
f	5441	o	11082	x	2361
g	3751	p	6891	y	3212
h	9868	q	1103	z	563
i	12827	r	10756		

The most frequent letter is e closely followed by t, n, and a. However, since this is a mathematical textbook, there is considerable distortion compared with a piece of prose caused by the extensive use of symbols, particularly in equations. What is the probability that a letter is i? What is the probability that a letter is a vowel a, e, i, o, u?
   A word count shows that the word *probability* occurs 821 times.

1.27 Using the table of probabilities in Example 1.9, calculate the conditional probabilities and random variable given by the conditional expectation $\mathbf{E}(Y|X)$.

1.28. (a) Show that the moment generating function (mgf) of a $N(\mu, \sigma^2)$ random variable is $\exp(\mu s + \frac{1}{2}\sigma^2 s^2)$.

---

[7] Augustin-Louis Cauchy (1789–1857), French mathematician.

(b) Use (a) to identify the distribution of $Z = (X - \mu)/\sigma$ where X is $N(\mu, \sigma^2)$.

(c) Use an mgf to identify the distribution of $Y = aX + b$ where a, b are constants and X is $N(\mu, \sigma^2)$.

(d) A sequence X_i $(i = 1, 2, \ldots n)$ of n independent normally distributed random variables has mean μ_i and variance σ_i^2. Derive the distribution of $\sum_{i=1}^{n} X_i$. If Z_i are n iid standard normal random variables, what is the distribution of $\sum_{i=1}^{n} Z_i$?

(e) Consider the sequence of random variables in (c): find the distribution of $\sum_{i=1}^{n} a_i X_i$, where the a_i's $(i = 1, 2, \ldots, n)$ are constants. What is the distribution of the average of the X_i's, namely $\sum_{i-1}^{n} X_i/n$?

1.29. (a) Let Z have a standard normal distribution. Show that the mgf of Z^2 is $(1 - 2s)^{-\frac{1}{2}}$. This is the mgf of a χ^2 distribution on 1 degree of freedom (χ_1^2). Hence Z^2 has a χ_1^2 distribution.

(b) Let $Z_1, Z_2 \ldots, Z_n$ be a sequence of n iid standard normal random variables. Show that the mgf of

$$Y = \sum_{i=1}^{n} Z_i^2$$

is $(1 - 2s)^{-\frac{1}{2}n}$, which is the mgf of a χ_n^2 distribution.

(c) Find the mean and variance of Y in (b).

CHAPTER 2

Some Gambling Problems

2.1 Gambler's ruin

Consider a game of chance between two players: A, the gambler and B, the opponent. It is assumed that at each play, A either wins one unit from B with probability p or loses one unit to B with probability $q = 1 - p$. Conversely, B either wins from A or loses to A with probabilities q or p. The result of every play of the game is independent of the results of previous plays. The gambler A and the opponent B each start with a given number of units and the game ends when either player has lost his or her initial stake. What is the probability that the gambler loses all his or her money or wins all the opponent's money, assuming that an unlimited number of plays are possible? This is the classic **gambler's ruin** problem[1]. In a simple example of gambler's ruin, each play could depend on the spin of a fair coin, in which case $p = q = \frac{1}{2}$. The word *ruin* is used because if the gambler plays a fair game against a bank or casino with unlimited funds, then the gambler is certain to lose.

   The problem will be solved by using results from conditional probability, which then leads to a **difference equation**, and we shall have more to say about methods of solution later. There are other questions associated with this problem, such as how many plays are expected before the game finishes. In some games the player might be playing against a casino which has a very large (effectively infinite) initial stake.

2.2 Probability of ruin

The result of each play of the game is a (modified) Bernoulli random variable (Section 1.7), which can only take the values -1 and $+1$. After a series of plays, we are interested in the current capital or stake of A, the gambler. This is simply the initial capital of A plus the sum of the values of the Bernoulli random variables generated by these plays. We are also interested in how the random variable which represents the current capital changes or evolves with the number of plays. This is measured at discrete points when the result of each play is known.

   Suppose that A has an initial capital of k units and B starts with $a - k$, where a and k are positive integers and $a > k$. If X_n is a random variable representing A's stake after n plays (or at time point n), then initially $X_0 = k$. If $X_n = 0$, then the

---

[1] The problem was first proposed by Pascal in 1656. Many mathematicians were attracted by this gambling question, including Fermat (1601–1665), Huygens (1629–1695) and others. For an account of the history of the problem see Song and Song (2013).

gambler A has lost (note that we must have $n \geq k$), whilst if $X_n = a$ $(n \geq a - k)$ then B is ruined, and in both cases the game terminates. Our initial objective is the derivation of $\mathbf{P}(X_n = 0)$ for all $n \geq k$.

The sequence of random variables $X_0, X_1, X_2 \ldots$ represents what is known as a **random process** with a finite sample space consisting of the integers from 0 to a. These values are known as the **state** of the process at each **stage** or time point n. If C_k is the event that A is eventually ruined when starting with initial capital k, then, by using the fact $X_n = 0$ $(n = k, k + 1, k + 2, \ldots)$, it follows that

$$\mathbf{P}(C_k) = \sum_{n=k}^{\infty} \mathbf{P}(X_n = 0).$$

Note again that the summation starts at $n = k$ since the minimum number of steps in which the game could end must be k. Note also that the results of each trial are independent, but X_n, $n = 0, 1, 2, \ldots$ are not. This is easily seen to be true by considering a particular value of X_n, say x, $(0 < x < a)$, after n plays, say. This event may only occur if previously $X_{n-1} = x - 1$ or $x + 1$. The state reached in any play depends on the state of the previous play only: in other words the process is said to display the **Markov** property, of which more will be explained later.

Clearly the calculation of $\mathbf{P}(X_n = 0)$ for all n is likely to be a long and tedious process. However, we now introduce a method for the calculation of these probabilities which avoids this: it is based on the solution of **linear homogeneous difference equations**.

Due to the sequential nature of this process, after the result of a play is known, then A's stake is either increased or decreased by one unit. This capital then becomes the new stake which, in turn, becomes the initial stake for the next play. Hence if we define $u_k = \mathbf{P}(C_k)$, then after the first play the probability of ruin is either $u_{k+1} = \mathbf{P}(C_{k+1})$ or $u_{k-1} = \mathbf{P}(C_{k-1})$. Let us consider the result of the first play, and define D to be the event that A wins, and the complement D^c the event that A loses. Using the law of total probability (Section 1.3), it follows that

$$\mathbf{P}(C_k) = \mathbf{P}(C_k|D)\mathbf{P}(D) + \mathbf{P}(C_k|D^c)\mathbf{P}(D^c). \tag{2.1}$$

As remarked previously, event C_k given a win, namely $C_k|D$ becomes event C_{k+1}. Hence $\mathbf{P}(C_k|D) = \mathbf{P}(C_{k+1})$. Similarly $\mathbf{P}(C_k|D^c) = \mathbf{P}(C_{k-1})$. Also $\mathbf{P}(D) = p$ and $\mathbf{P}(D^c) = q$, which means that Eqn (2.1) can be written as

$$u_k = u_{k+1}p + u_{k-1}q, \qquad (1 \leq k \leq a - 1).$$

This equation can be re-arranged into

$$pu_{k+1} - u_k + qu_{k-1} = 0, \tag{2.2}$$

which is a second-order linear homogeneous difference equation. It is described as homogeneous since there is no term on the right-hand side of Eqn (2.2), and second-order since the sequence difference between u_{k+1} and u_{k-1} is $(k+1) - (k-1) = 2$. If the gambler starts with zero stake, then ruin is certain, whilst if the gambler starts with all the capital a, then ruin is impossible. These translate into

$$u_0 = \mathbf{P}(C_0) = 1 \qquad \text{and} \qquad u_a = \mathbf{P}(C_a) = 0, \tag{2.3}$$

which are the **boundary conditions** for the difference equation (2.2).

It is also possible to solve the equation by iteration (see Tuckwell (1995)): this will be the subject of Problem 2.17. However here we shall describe a method which also has a more general application (see Jordan and Smith (2008) for a more detailed explanation of the solution of difference equations). Consider a solution of the form

$$u_k = m^k,$$

where m is to be determined. Direct substitution into the left-hand side of Eqn (2.2) yields

$$pu_{k+1} - u_k + qu_{k-1} = m^{k-1}[pm^2 - m + q].$$

This is zero if either $m = 0$, which is known as the trivial solution and is not usually of interest in this context, or if m satisfies the quadratic equation

$$pm^2 - m + q = (pm - q)(m - 1) = 0, \qquad (p + q = 1),$$

which is known as the **characteristic equation** of the difference equation (2.2). The roots of the equation are $m_1 = 1$ and $m_2 = q/p$. Since the difference equation is linear, the general solution is, provided that $p \neq q$, any linear combination of the two solutions with $m = m_1$ and $m = m_2$, that is

$$u_k = A_1 m_1^k + A_2 m_2^k = A_1 + A_2 \left(\frac{q}{p} \right)^k,$$

where A_1 and A_2 are arbitrary constants. The boundary conditions $u_0 = 1$ and $u_a = 0$ imply that

$$A_1 + A_2 = 1 \qquad \text{and} \qquad A_1 + A_2 s^a = 0,$$

so that

$$A_1 = -\frac{s^a}{1 - s^a} \qquad \text{and} \qquad A_2 = \frac{1}{1 - s^a},$$

where $s = q/p$. Hence the probability that the gambler is ruined given an initial capital of k is, if $p \neq \frac{1}{2}$,

$$u_k = \frac{s^k - s^a}{1 - s^a}. \tag{2.4}$$

The special case $p = q = \frac{1}{2}$ has to be treated separately. The characteristic equation of

$$\tfrac{1}{2} u_{k+1} - u_k + \tfrac{1}{2} u_{k-1} = 0$$

is

$$m^2 - 2m + 1 = 0,$$

which has the repeated root $m_1 = m_2 = 1$. In this case one solution is 1, but we still require a second independent solution. For a repeated root of difference equations we try $k \times$ (the repeated root). Hence

$$\tfrac{1}{2} u_{k+1} - u_k + \tfrac{1}{2} u_{k-1} = \tfrac{1}{2}(k + 1) - k + \tfrac{1}{2}(k - 1) = 0,$$

implying that $u_k = k$ is a second independent solution. Thus the general solution is

$$u_k = A_1 + A_2 k.$$

With the same boundary conditions, it follows that

$$u_k = \frac{a - k}{a}. \tag{2.5}$$

If the game is now looked at from B's viewpoint, then to obtain his/her probability of ruin, v_{a-k}, say, there is no need to derive this from first principles. In the above derivation, simply interchange p and q, and replace k by $a - k$ in the results (2.4) and (2.5). Hence B is ruined with probability

$$v_{u-k} = \left(\frac{1}{s^{a-k}} - \frac{1}{s^a} \right) \bigg/ \left(1 - \frac{1}{s^a} \right) = \frac{s^k - 1}{s^a - 1},$$

if $s \neq 1$, and with probability

$$v_{a-k} = \frac{k}{a},$$

if $s = 1$.

It follows that

$$u_k + v_{a-k} = 1$$

in both cases. Hence the game must terminate eventually with one of the players losing.

Example 2.1. *Suppose the possibility of a draw is now included. Let the probability that the gambler wins, loses, or draws against the opponent in a play be respectively p, q, or r. We assume that these are the only possible outcomes so that $p + q + r = 1$. Show that the probability of ruin u_k is given by*

$$u_k = \frac{s^k - s^a}{1 - s^a}, \qquad s = \frac{q}{p} \neq 1, \quad (p \neq \tfrac{1}{2}),$$

where the gambler's initial stake is k and the total at stake is a.

In this case the law of total probability given by (2.1) is extended to include a third possible outcome. In the draw the stakes remain unchanged after the play. Hence the probability of ruin u_k satisfies

$$u_k = u_{k+1}p + u_k r + u_{k-1}q,$$

or

$$pu_{k+1} - (1 - r)u_k + qu_{k-1} = 0. \tag{2.6}$$

As in the main problem, the boundary conditions are $u_0 = 1$ and $u_a = 0$. Replace $1 - r$ by $p + q$ in the difference equation. Its characteristic equation becomes

$$pm^2 - (p + q)m + q = (pm - q)(m - 1) = 0,$$

which has the general solution

$$u_k = A_1 + A_2 \left(\frac{q}{p} \right)^k.$$

Notice that this is the same solution as the standard problem with no draws; q/p is still the ratio of the probability of losing to winning at each play. The only difference is that $p + q \neq 1$ in this case. Hence as in (2.4)

$$u_k = \frac{s^k - s^a}{1 - s^a}, \tag{2.7}$$

where $s = q/p$.

Example 2.2. *In a gambler's ruin game the initial stakes are k and $a - k$ ($1 \leq k \leq a - 2$) for the two players: if the gambler wins then she or he wins two units but if the gambler loses then she or he loses one unit. If at some point in the game the gambler has $a - 1$ units then only one unit can be won. The probability of win or lose at each play is $p = \frac{1}{3}$ or $q = \frac{2}{3}$. What are the boundary conditions and what is the probability of ruin?*

If u_k is the probability of the gambler's ruin, then the law of total probability has to be changed in this problem to

$$u_k = pu_{k+2} + (1-p)u_{k-1},$$

for $1 \leq k \leq a - 2$.

The boundary conditions $u_0 = 1$ and $u_a = 0$ still hold. If $k = a - 1$, then the law of total probability has to be amended to

$$u_{a-1} = pu_a + (1-p)u_{a-2},$$

or

$$u_{a-1} - (1-p)u_{a-2} = 0,$$

and this becomes the third boundary condition. Three boundary conditions are required since the difference equation for u_k is now the *third-order* equation

$$pu_{k+2} - u_k + (1-p)u_{k-1} = 0.$$

Its characteristic equation is, with $p = \frac{1}{3}$,

$$m^3 - 3m + 2 = (m-1)^2(m+2) = 0,$$

which has the roots $1, 1, -2$. Hence the general solution is (note the solution for the repeated root)

$$u_k = A_1 + A_2 k + A_3(-2)^k. \tag{2.8}$$

The three boundary conditions above lead to

$$A_1 + A_3 = 1,$$

$$A_1 + A_2 a + A_3(-2)^a = 0,$$

$$A_1 + A_2(a-1) + A_3(-2)^{a-1} = \tfrac{2}{3}(A_1 + A_2(a-2) + A_3(-2)^{a-2}),$$

for A_1, A_2, and A_3. The solutions are

$$A_1 = 1 - A_3 = \frac{(1+3a)(-2)^a}{(1+3a)(-2)^a - 1}, \qquad A_2 = \frac{-3(-2)^a}{(1+3a)(-2)^a - 1},$$

from which it follows that

$$u_k = \frac{(1+3a-3k)(-2)^a - (-2)^k}{(1+3a)(-2)^a - 1}.$$

2.3 Some numerical simulations

It is very easy now with numerical and symbolic computation to simulate probability problems with simple programs. Figure 2.1 shows a simulation of the gambler's ruin in the case $a = 20$, $k = 10$, and $p = \frac{1}{2}$: the theoretical probability of ruin is given by Eqn (2.5), namely $u_k = (a-k)/a = \frac{1}{2}$. The program runs until the gambler's stake

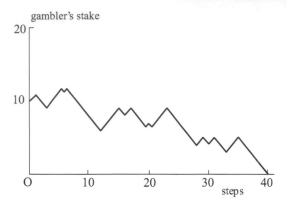

Figure 2.1 *The figure shows the steps in a simulation of a gambler's ruin problem with $a = 20$ and $k = 10$ with probability $p = \frac{1}{2}$ at each play. Ruin occurs after 40 steps in this case, although by (2.5) the gambler has even chances of winning or losing the game.*

is either 0 or 20 in this example. Some sample probabilities computed from Eqn (2.4) and Eqn (2.5):

$$u_k = \frac{s^k - s^a}{1 - s^a}, \quad (p \neq \tfrac{1}{2}), \quad u_k = \frac{a - k}{a}, \quad (p = \tfrac{1}{2}),$$

are shown in Figure 2.2, again with $a = 20$ for various initial probabilities and play probabilities p. So, for example, the probability of ruin with $p = 0.42$ and initial

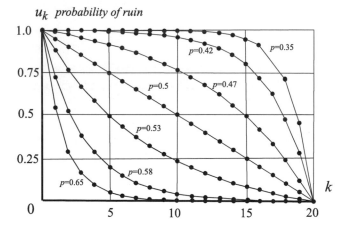

Figure 2.2 *The ruin probability $u_k = (s^k - s^a)/(1 - s^a)$, $(p \neq \frac{1}{2})$, $u_k = (a - k)/a$, $(p = \frac{1}{2})$ versus k for $a = 20$ and a sample of probabilities p.*

stake of 17 is about 0.62. The figure shows that, in a game in which both parties start with the same initial stake 10, the gambler's probability of ruin becomes very close to certainty for any p less than 0.35. An alternative view is shown in Figure 2.3, which shows the value of u_{10} versus p. The figure emphasizes that there is only a real

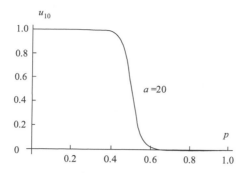

Figure 2.3 *The probability u_{10} is shown against the play probability p for the case $a = 20$.*

contest between the players if p lies between 0.4 and 0.6.

2.4 Duration of the game

It is natural in the gambler's ruin problem to be curious about how long, or really how many plays we would expect the game to last. This is the number of steps to termination in which either the gambler or the opponent loses the game. Let us first consider a situation where the state of the game is defined in terms of two variables: k, the initial capital, and n, the remaining number of plays until the end of the game. Now n is unknown, and is a value of the random variable N which depends, in turn, on the results of the remaining plays.

Let $p(n|k)$ be the conditional probability that the game ends in n steps given that the initial capital is k. Clearly n will be any positive integer greater than or equal to the smaller of k and $a - k$ since if the gambler won (lost) *every* play then s/he would win (lose) the game in $a - k$ (k) plays. Let the random variable N be the number of plays until the game ends, and let K be the random variable of the initial stake. The expected number of plays to termination, or, as it is also known, the **expected duration**, will be the *conditional expectation* and random variable

$$\mathbf{E}(N|K) = \sum_{n=0}^{\infty} np(n|k) = d_k, \qquad (2.9)$$

say (see Section 1.10 for discussion of conditional expectation). We have proved in Section 2.2 that termination is certain eventually so that $p(n|k)$ is a probability function and must therefore satisfy

$$\sum_{n=0}^{\infty} p(n|k) = 1,$$

for each fixed k. After the result of the next play is known, then the process will move from step (k, n) to either step $(k + 1, n - 1)$ with probability p, or to step $(k - 1, n - 1)$ with probability $q = 1 - p$. By the law of total probability, it follows that

$$p(n|k) = p(n - 1|k + 1)p + p(n - 1|k - 1)q, \qquad (n, k \geq 1).$$

Substituting for $p(n|k)$ in Eqn (2.9), we obtain the expected duration d_k given by

$$d_k = p \sum_{n=1}^{\infty} np(n-1|k+1) + q \sum_{n=1}^{\infty} np(n-1|k-1).$$

The change of variable $r = n - 1$ in both summations leads to

$$
\begin{aligned}
d_k &= p \sum_{r=0}^{\infty} (r+1)p(r|k+1) + q \sum_{r=0}^{\infty} (r+1)p(r|k-1), \\
&= p \sum_{r=1}^{\infty} rp(r|k+1) + q \sum_{r=1}^{\infty} rp(r|k-1) + p \sum_{r=0}^{\infty} p(r|k+1) + \\
&\quad\; q \sum_{r=0}^{\infty} p(r|k-1), \\
&= p \sum_{r=1}^{\infty} rp(r|k+1) + q \sum_{r=1}^{\infty} rp(r|k-1) + p + q,
\end{aligned}
$$

since for the probability functions $p(r|k+1)$ and $p(r|k-1)$,

$$\sum_{r=0}^{\infty} p_{k+1}(r|k+1) = \sum_{r=0}^{\infty} p_{k-1}(r|k-1) = 1.$$

Hence the expected duration d_k satisfies the difference equation

$$d_k = pd_{k+1} + qd_{k-1} + 1, \qquad (k \geq 1)$$

since $p + q = 1$ and

$$d_{k+1} = \sum_{r=1}^{\infty} rp(r|k+1) \quad \text{and} \quad d_{k-1} = \sum_{r=1}^{\infty} rp(r|k-1).$$

This equation can be re-arranged into

$$pd_{k+1} - d_k + qd_{k-1} = -1, \tag{2.10}$$

which is similar to the difference equation for the probability u_k except for the term on the right-hand side. This is a linear inhomogeneous second-order difference equation. The boundary conditions are again obtained by considering the extremes where one of the players loses. Thus if $k = 0$ or $k = a$, then the game terminates so that the expected durations must be zero, that is,

$$d_0 = d_a = 0.$$

The solution to this type of difference equation is the sum of the general solution of the corresponding *homogeneous* equation, known as the **complementary function**, and a **particular solution** of the complete equation. The homogeneous equation has, for $s = q/p \neq 1$, the general solution

$$A_1 + A_2 s^k$$

as in Eqn (2.2). For the particular solution we look at the right-hand side and try a suitable function. Since any constant is a solution of the homogeneous equation, it cannot also be a solution of the inhomogeneous equation. Instead, we try $d_k = Ck$, where C is a constant. Thus

$$pd_{k+1} - d_k + qd_{k-1} + 1 = pC(k+1) - Ck + qC(k-1) + 1 = C(p-q) + 1 = 0,$$

for all k if $C = 1/(q-p)$. Hence the full general solution is

$$d_k = A_1 + A_2 s^k + \frac{k}{q-p}.$$

The boundary conditions imply

$$A_1 + A_2 = 0, \qquad \text{and} \qquad A_1 + A_2 s^a + \frac{a}{q-p} = 0.$$

Hence

$$A_1 = -A_2 = -\frac{a}{(q-p)(1-s^a)},$$

with the result that the expected duration for $s \neq 1$ is

$$d_k = -\frac{a(1-s^k)}{(q-p)(1-s^a)} + \frac{k}{q-p} = \frac{1}{1-2p}\left[k - \frac{a(1-s^k)}{1-s^a}\right]. \qquad (2.11)$$

If $s = 1$, the difference equation (2.12) becomes

$$d_{k+1} - 2d_k + d_{k-1} = -2.$$

The general solution of the homogeneous equation is

$$A_1 + A_2 k,$$

which means that neither C nor Ck can satisfy the inhomogeneous equation. Instead we try Ck^2 for the particular solution. Thus

$$d_{k+1} - 2d_k + d_{k-1} + 2 = C(k+1)^2 - 2Ck^2 + C(k-1)^2 + 2 = 2C + 2 = 0,$$

if $C = -1$. Hence

$$d_k = A_1 + A_2 k - k^2 = k(a-k).$$

The boundary conditions imply $A_1 = 0$ and $A_2 = a$, so that

$$d_k = k(a-k). \qquad (2.12)$$

A sample of expected durations are shown in Figure 2.4 for a total stake of $a = 20$ and different probabilities p. For $p = \frac{1}{2}$ the expected duration has a maximum number of 100 when $k = 10$. Hence, on average in this case, a game in which both players start with 10 units each will last for 100 plays. Generally if $a = 2k$, each player starting with a stake of k and $p = \frac{1}{2}$, then the expected duration behaves as k^2.

2.5 Some variations of gambler's ruin

2.5.1 The infinitely rich opponent

Consider the gambler's ruin problem in which the opponent is assumed to be infinitely rich. This models a gambler playing against a bank or a casino whose re-

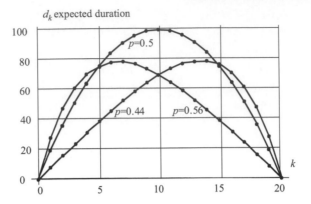

Figure 2.4 *Expected duration d_k against k for $a = 20$ and a selection of play probabilities p.*

sources are very large. As before, the gambler's initial stake is a finite integer k, but effectively the bank's resources are infinite, that is, $a = \infty$. We approach the problem by looking at the finite case, and then find the limits of u_k and the expected duration, d_k, as $a \to \infty$. As we might expect, the results depend on magnitude of $s = q/p$, the ratio of the probabilities.

   (a) $s = q/p < 1$, $(p > \frac{1}{2})$. The gambler A has the advantage in each play. Since $\lim_{a \to \infty} s^a = 0$, it follows from Eqn (2.4) that

$$\lim u_k = \lim_{a \to \infty} \frac{s^k - s^a}{1 - s^a} = s^k < 1.$$

The expected number of plays until this happens is given by Eqn (2.11), which states that

$$d_k = -\frac{a(1 - s^k)}{(q - p)(1 - s^a)} + \frac{k}{q - p}.$$

Again $s^a \to 0$, since $s < 1$, so that a in the first term dominates with the result that $\lim_{a \to \infty} d_k = \infty$. Hence A is not certain to be ruined but the game would be expected to take an infinite number of plays, which itself is to be anticipated since B has an infinite stake.

   (b) $s > 1$, $(p < \frac{1}{2})$. B has the advantage in each play. In this case $\lim_{a \to \infty} s^a = \infty$, so that, from Eqn (2.4),

$$\lim_{a \to \infty} u_k = \lim_{a \to \infty} \frac{(s^k/s^a) - 1}{(1/s^a) - 1} = 1,$$

and

$$\lim_{a \to \infty} d_k = \lim_{a \to \infty} \frac{-a}{s^a} \frac{(1 - s^k)}{(\frac{1}{s^a} - 1)(q - p)} + \frac{k}{q - p} = \frac{k}{q - p},$$

since

$$\lim_{a \to \infty} \frac{a}{s^a} = 0.$$

As might be anticipated, ruin is certain and the expected duration is finite.

(c) $s = 1$ or $p = \frac{1}{2}$. For the case of equal probabilities,

$$\lim_{a \to \infty} u_k = \lim_{a \to \infty} \left(1 - \frac{k}{a}\right) = 1, \quad \text{and} \quad \lim_{a \to \infty} d_k = \lim_{a \to \infty} k(a - k) = \infty.$$

Ruin is certain but it may take a great deal of time.

2.5.2 The generous opponent

Suppose that both players start with finite capital, and suppose that whenever A loses his/her last unit, one unit is returned to A so that A is never ruined. The opponent B is the generous gambler. As a consequence we might expect that B must be ruined since A cannot lose. Hence there can be no possibility, other than $u_k = 0$. This can be checked by solving Eqn (2.2) subject to the boundary conditions

$$u_0 = u_1, \text{ and } u_a = 0.$$

For the expected duration the boundary condition at $k = 0$ must be modified. Since one unit is returned, the expected duration at $k = 0$ must be the same as that at $k = 1$. The boundary conditions become

$$d_0 = d_1, \quad \text{and} \quad d_a = 0.$$

Consider the case of equal probabilities at each play. Then for $p = \frac{1}{2}$,

$$d_k = A_1 + A_2 k - k^2.$$

Hence $A_1 = a^2 - a$ and $A_2 = 1$, so that

$$d_k = (a - k)(a + k - 1).$$

There are thus $(a - k)(a - 1)$ more plays than in the standard game.

2.5.3 Changing the stakes

Suppose that in the original game the stakes per play are halved for both players so that in the new game A has effectively $2k$ units and B has $2(a - k)$ units. How is the probability of ruin changed?

Let v_k be the probability of ruin. Then, by analogy with the formula for u_k given by Eqn (2.4),

$$v_k = \frac{s^{2k} - s^{2a}}{1 - s^{2a}} = \frac{(s^k - s^a)(s^k + s^a)}{(1 - s^a)(1 + s^a)} = u_k \frac{s^k - s^a}{1 + s^a}, \quad s \neq 1.$$

If $s < 1$, then

$$\frac{s^k + s^a}{1 + s^a} < 1$$

so that $v_k < u_k$. On the other hand, if $s > 1$, then

$$\frac{s^k + s^a}{1 + s^a} > 1.$$

so that $v_k > u_k$. If $s < 1$ namely $p > \frac{1}{2}$, then it could be wise for A to agree to this change of play. As might be expected for $s = 1$, that is, equal probabilities with $p = \frac{1}{2}$, the probability of ruin is unaffected.

2.6 Problems

2.1. In the standard gambler's ruin problem, with total stake a and gambler's stake k, and the gambler's probability of winning at each play is p, calculate the probability of ruin in the following cases;
(a) $a = 100, k = 5, p = 0.6$;
(b) $a = 80, k = 70, p = 0.45$;
(c) $a = 50, k = 40, p = 0.5$.
Also find the expected duration in each case.

2.2. In a casino game based on the standard gambler's ruin, the gambler and the dealer each start with 20 tokens and one token is bet on at each play. The game continues until one player has no further tokens. It is decreed that the probability that any gambler is ruined is 0.52 to protect the casino's profit. What should the probability that the gambler wins at each play be?

2.3. Find general solutions of the following difference equations:
(a) $u_{k+1} - 4u_k + 3u_{k-1} = 0$;
(b) $7u_{k+2} - 8u_{k+1} + u_k = 0$;
(c) $u_{k+1} - 3u_k + u_{k-1} + u_{k-2} = 0$;
(d) $pu_{k+2} - u_k + (1-p)u_{k-1} = 0, \quad (0 < p < 1)$.

2.4 Solve the following difference equations subject to the given boundary conditions:
(a) $u_{k+1} - 6u_k + 5u_{k-1} = 0, \quad u_0 = 1, u_4 = 0$;
(b) $u_{k+1} - 2u_k + u_{k-1} = 0, \quad u_0 = 1, u_{20} = 0$;
(c) $d_{k+1} - 2d_k + d_{k-1} = -2, \quad d_0 = 0, d_{10} = 0$;
(d) $u_{k+2} - 3u_k + 2u_{k-1} = 0, \quad u_0 = 1, u_{10} = 0, 3u_9 = 2u_8$.

2.5. Show that a difference equation of the form

$$au_{k+2} + bu_{k+1} - u_k + cu_{k-1} = 0,$$

where $a, b, c \geq 0$ are probabilities with $a + b + c = 1$, can never have a characteristic equation with complex roots.

2.6. In the standard gambler's ruin problem with equal probabilities $p = q = \frac{1}{2}$, find the expected duration of the game given the usual initial stakes of k units for the gambler and $a - k$ units for the opponent.

2.7. In a gambler's ruin problem the possibility of a draw is included. Let the probability that the gambler wins, loses, or draws against an opponent be, respectively, $p, p, 1 - 2p, (0 < p < \frac{1}{2})$. Find the probability that the gambler loses the game, given the usual initial stakes of k units for the gambler and $a - k$ units for the opponent. Show that d_k, the expected duration of the game, satisfies

$$pd_{k+1} - 2pd_k + pd_{k-1} = -1.$$

Solve the difference equation and hence find the expected duration of the game.

2.8. In the changing stakes game in which a game is replayed with each player having twice as many units, $2k$ and $2(a-k)$ respectively, suppose that the probability of a win for the gambler at each play is $\frac{1}{2}$. Whilst the probability of ruin is unaffected, by how much is the expected duration of the game extended compared with the original game?

2.9. A roulette wheel has 37 radial slots of which 18 are red, 18 are black, and 1 is green. The gambler bets one unit on either red or black. If the ball falls into a slot of the same color, then the gambler wins one unit, and if the ball falls into the other color (red or black), then the casino wins. If the ball lands in the green slot, then the bet remains for the next spin of the wheel or more if necessary until the ball lands on a red or black slot. The original bet is either *returned* or *lost* depending on whether the outcome matches the original bet or not (this is the Monte Carlo system). Show that the probability u_k of ruin for a gambler who starts with k chips with the casino holding $a - k$ chips satisfies the difference equation

$$36u_{k+1} - 73u_k + 37u_{k-1} = 0.$$

Solve the difference equation for u_k. If the house starts with 1,000,000 euros at the roulette wheel and the gambler starts with 10,000 euros, what is the probability that the gambler breaks the bank if 5,000 euros are bet at each play?

In the US system the rules are less generous to the players. If the ball lands on green then the player simply loses. What is the probability now that the player wins given the same initial stakes? (See Luenberger (1979).)

2.10. In a single trial the possible scores 1 and 2 can each occur with probability $\frac{1}{2}$. If p_m is the probability of scoring *exactly* m points at some stage, that is the score after several trials is the sum of individual scores in each trial. Show that

$$p_m - \tfrac{1}{2}p_{m-1} + \tfrac{1}{2}p_{m-2}.$$

Calculate p_1 and p_2, and find a formula for p_m. How does p_m behave as m becomes large? How do you interpret the result?

2.11. In a single trial the possible scores 1 and 2 can occur with probabilities q and $1 - q$, where $0 < q < 1$. Find the probability of scoring exactly n points at some stage in an indefinite succession of trials. Show that

$$p_n \to \frac{1}{2 - q},$$

as $n \to \infty$.

2.12. The probability of success in a single trial is $\frac{1}{3}$. If u_k is the probability that there are no two consecutive successes in k trials, show that u_k satisfies

$$u_{k+1} = \tfrac{2}{3}u_k + \tfrac{2}{9}u_{k-1}.$$

What are the values of u_1 and u_2? Hence show that

$$u_k = \frac{1}{6}\left[(3 + 2\sqrt{3})\left(\frac{1 + \sqrt{3}}{3}\right)^k + (3 - 2\sqrt{3})\left(\frac{1 - \sqrt{3}}{3}\right)^k\right].$$

2.13. A gambler with initial capital k units plays against an opponent with initial capital $a -

k units. At each play of the game the gambler either wins one unit or loses one unit with probability $\frac{1}{2}$. Whenever the opponent loses the game, the gambler returns one unit so that the game may continue. Show that the expected duration of the game is $k(2a - 1 - k)$ plays.

2.14. In the usual gambler's ruin problem, the probability that the gambler is eventually ruined is

$$u_k = \frac{s^k - s^a}{1 - s^a}, \qquad s = \frac{q}{p}, \qquad (p \neq \tfrac{1}{2}).$$

In a new game the stakes are halved, whilst the players start with the same initial sums. How does this affect the probability of losing for the gambler? Should the gambler agree to this change of rule if $p < \frac{1}{2}$? By how many plays is the expected duration of the game extended?

2.15. In a gambler's ruin game, suppose that the gambler can win £2 with probability $\frac{1}{3}$ or lose £1 with probability $\frac{2}{3}$. Show that

$$u_k = \frac{(3k - 1 - 3a)(-2)^a + (-2)^k}{1 - (3a + 1)(-2)^a}.$$

Compute u_k if $a = 9$ for $k = 1, 2, \ldots, 8$.

2.16. Find the general solution of the difference equation

$$u_{k+2} - 3u_k + 2u_{k-1} = 0.$$

A reservoir with total capacity of a volume units of water has, during each day, either a net inflow of two units with probability $\frac{1}{3}$ or a net outflow of one unit with probability $\frac{2}{3}$. If the reservoir is full or nearly full, any excess inflow is lost in an overflow. Derive a difference equation for this model for u_k, the probability that the reservoir will eventually become empty given that it initially contains k units. Explain why the upper boundary conditions can be written $u_a = u_{a-1}$ and $u_a = u_{a-2}$. Show that the reservoir is certain to be empty at some time in the future.

2.17. Consider the standard gambler's ruin problem in which the total stake is a and gambler's stake is k, and the gambler's probability of winning at each play is p and losing is $q = 1 - p$. Find u_k, the probability of the gambler losing the game, by the following alternative method. List the difference Eqn (2.2) as

$$\begin{aligned}
u_2 - u_1 &= s(u_1 - u_0) = s(u_1 - 1) \\
u_3 - u_2 &= s(u_2 - u_1) = s^2(u_1 - 1) \\
&\vdots \\
u_k - u_{k-1} &= s(u_{k-1} - u_{k-2}) = s^{k-1}(u_1 - 1),
\end{aligned}$$

where $s = q/p \neq \frac{1}{2}$ and $k = 2, 3, \ldots a$. The boundary condition $u_0 = 1$ has been used in the first equation. By adding the equations, show that

$$u_k = u_1 + (u_1 - 1)\frac{s - s^k}{1 - s}.$$

Determine u_1 from the other boundary condition $u_a = 0$, and hence find u_k. Adapt the same method for the special case $p = q = \frac{1}{2}$.

2.18. A car park has 100 parking spaces. Cars arrive and leave randomly. Arrivals or departures of cars are equally likely, and it is assumed that simultaneous events have negligible probability. The 'state' of the car park changes whenever a car arrives or departs. Given that at some instant there are k cars in the car park, let u_k be the probability that the car park first becomes full before it becomes empty. What are the boundary conditions for u_0 and u_{100}? How many car movements can be expected before this occurs?

2.19. In a standard gambler's ruin problem with the usual parameters, the probability that the gambler loses is given by

$$u_k = \frac{s^k - s^a}{1 - s^a}, \qquad s = \frac{1 - p}{p}.$$

If p is close to $\frac{1}{2}$, given say by $p = \frac{1}{2} + \varepsilon$ where $|\varepsilon|$ is small, show, by using binomial expansions, that

$$u_k = \frac{a - k}{a} \left[1 - 2k\varepsilon - \frac{4}{3}(a - 2k)\varepsilon^2 + O(\varepsilon^3) \right]$$

as $\varepsilon \to 0$. (The order O terminology is defined as follows: we say that a function $g(\varepsilon) = O(\varepsilon^b)$ as $\varepsilon \to 0$ if $g(\varepsilon)/\varepsilon^b$ is bounded in a neighborhood which contains $\varepsilon = 0$.)

2.20. A gambler plays a game against a casino according to the following rules. The gambler and casino each start with 10 chips. From a deck of 53 playing cards which includes a joker, cards are randomly and successively drawn with replacement. If the card is red or the joker, the casino wins 1 chip from the gambler, and if the card is black the gambler wins 1 chip from the casino. The game continues until either player has no chips. What is the probability that the gambler wins? What will be the expected duration of the game?

2.21. In the standard gambler's ruin problem with total stake a and gambler's stake k, the probability that the gambler loses is

$$u_k = \frac{s^k - s^a}{1 - s^a},$$

where $s = (1 - p)/p$. Suppose that $u_k = \frac{1}{2}$, that is, fair odds. Express k as a function of a. Show that

$$k = \frac{\ln[\frac{1}{2}(1 + s^a)]}{\ln s}.$$

Of course this value of k can only be an approximation since it is generally not an integer.

2.22. In a gambler's ruin game the probability that the gambler wins at each play is α_k and loses is $1 - \alpha_k$, $(0 < \alpha_k < 1, \ 0 \le k \le a - 1)$, that is, the probability varies with the current stake. The probability u_k that the gambler eventually loses satisfies

$$u_k = \alpha_k u_{k+1} + (1 - \alpha_k)u_{k-1}, \qquad u_o = 1, \qquad u_a = 0.$$

Suppose that u_k is a specified function such that $0 < u_k < 1$, $(1 \le k \le a - 1)$, $u_0 = 1$, and $u_a = 0$. Express α_k in terms of u_{k-1}, u_k, and u_{k+1}.
   Find α_k in the following cases:
(a) $u_k = (a - k)/a$;
(b) $u_k = (a^2 - k^2)/a^2$;
(c) $u_k = 1/(a + k)$.

2.23. In a gambler's ruin game the probability that the gambler wins at each play is α_k and loses is $1 - \alpha_k$, $(0 < \alpha_k < 1,\ 1 \le k \le a - 1)$, that is, the probability varies with the current stake. The probability u_k that the gambler eventually loses satisfies

$$u_k = \alpha_k u_{k+1} + (1 - \alpha_k) u_{k-1}, \quad u_o = 1, \quad u_a = 0.$$

Reformulate the difference equation as

$$u_{k+1} - u_k = \beta_k (u_k - u_{k-1}),$$

where $\beta_k = (1 - \alpha_k)/\alpha_k$. Hence show that

$$u_k = u_1 + \gamma_{k-1}(u_1 - 1), \quad (k = 2, 3, \ldots, a)$$

where

$$\gamma_k = \beta_1 + \beta_1 \beta_2 + \cdots + \beta_1 \beta_2 \ldots \beta_k.$$

Using the boundary condition at $k = a$, confirm that

$$u_k = \frac{\gamma_{a-1} - \gamma_{k-1}}{1 + \gamma_{a-1}}.$$

Check that this formula gives the usual answer if $\alpha_k = p \ne \frac{1}{2}$, a constant.

2.24. Suppose that a fair n-sided die is rolled n independent times. A match is said to occur if side i is observed on the ith trial, where $i = 1, 2, \ldots, n$.
(a) Show that the probability of at least one match is

$$1 - \left(1 - \frac{1}{n}\right)^n.$$

(b) What is the limit of this probability as $n \to \infty$?
(c) What is the probability that just one match occurs in n trials?
(d) What value does this probability approach as $n \to \infty$?
(e) What is the probability that two or more matches occur in n trials?

2.25. (Kelly's[2] strategy) A gambler plays a repeated favourable game in which the gambler wins with probability $p > \frac{1}{2}$ and loses with probability $q = 1 - p$. The gambler starts with an initial outlay K_0 (in some currency). For the first game the player bets a proportion rK_0, $(0 < r < 1)$. Hence, after this play the stake is $K_0(1 + r)$ after a win or $K_0(1 - r)$ after losing. Subsequently, the gambler bets the same proportion of the current stake at each play. Hence, after n plays of which w_n are wins the stake S_r will be

$$K_n(r) = K_0(1 + r)^{w_n}(1 - r)^{n - w_n}.$$

Construct the function

$$G_n(r) = \frac{1}{n} \ln \left[\frac{K_n(r)}{K_0}\right].$$

What is the expected value of $G_n(r)$ for *large* n? For what values of r is this expected value a maximum? This value of r indicates a safe betting level to maximise the gain, although at a slow rate. You might consider why the logarithm is chosen: this is known as a **utility function** in gambling and economics. It is a matter of choice and is a balance between having a reasonable gain against having a high risk gain. Calculate r if $p = 0.55$. [At the extremes $r = 0$ corresponds to no bet whilst $r = 1$ corresponds to betting K_0—the whole stake—in one go, which could be catastrophic.]

[2] John L. Kelly (1923–1965), American scientist.

CHAPTER 3

Random Walks

3.1 Introduction

Another way of modelling the gambler's ruin problem of the last chapter is as a one-dimensional **random walk**. Suppose that $a + 1$ positions are marked out on a straight line and numbered $0, 1, 2, \ldots, a$. A person starts at k where $0 < k < a$. The walk proceeds in such a way that at each step there is a probability p that the walker goes 'forward' one place to $k + 1$, and a probability $q = 1 - p$ that the walker goes 'back' one place to $k - 1$. The walk continues until either 0 or a is reached, and then ends. Generally, in a random walk, the position of a walker after having moved n times is known as the **state** of the walk after n **steps** or after covering n **stages**. Thus the walk described above starts at stage k at step 0 and moves to either stage $k - 1$ or stage $k + 1$ after 1 step, and so on. A random walk is said to be **symmetric** if $p = q = \frac{1}{2}$.

If the walk is bounded, then the ends of the walk are known as **barriers**, and they may have various properties. In this case the barriers are said to be **absorbing**, which implies that the walk must end once a barrier is reached since there is no escape. On the other hand, the barrier could be **reflecting**, in which case the walk returns to its previous state. A useful diagrammatic way of representing random walks is by a **transition** or **process diagram** as shown in Figure 3.1. In a transition diagram the possible stages of the walker can be represented by points on a line. If a transition between two points can occur in one step, then those points are joined by a curve or **edge**, as shown with an arrow indicating the direction of the walk and a *weighting* denoting the probability of the step occurring. In discrete mathematics or graph theory the transition diagram is known as a **directed graph**. A walk in the transition diagram is a succession of edges covered without a break. In Figure 3.1 the closed loops with weightings of 1 at the ends of the walk indicate the absorbing barriers with no escape.

Figure 3.1 *Transition diagram for a random walk with absorbing barriers at each end of the walk.*

3.2 Unrestricted random walks

A simple random walk on a line or in one dimension occurs when a step forward
$(+1)$ has probability p and a step back (-1) has probability $q(= 1 - p)$. At the i-th
step the modified Bernoulli random variable W_i (see Section 1.7) is observed, and
the position of the walk at the n-th step is the random variable:

$$X_n = X_0 + \sum_{i=1}^{n} W_i = X_{n-1} + W_n. \tag{3.1}$$

In the gambler's ruin problem, $X_0 = k$, but in the following discussion it is assumed,
without loss of generality, that walks start from the origin so that $X_0 = 0$.

The random walks described so far are restricted by barriers. We now consider
random walks without barriers, or **unrestricted random walks** as they are known. In
these walks, the position or state x can take any of the values $\{\ldots, -2, -1, 0, 1, 2, \ldots\}$.
In particular, we are interested in the position of the walk after a number of steps and
the probability of a return to the origin, the start of the walk. As seen from Eqn (3.1),
the position of the walk at step n simply depends on the position at the $(n-1)$th step.
This means that the simple random walk possesses what is known as the **Markov
property**: the current state of the walk depends on its immediate previous state, not
on the history of the walk up to the present state. Furthermore $X_n = X_{n-1} \pm 1$, and
we know that the transition probabilities from one position to another

$$\mathbf{P}(X_n = j | X_{n-1} = j - 1) = p,$$

and

$$\mathbf{P}(X_n = j | X_{n-1} = j + 1) = q,$$

are independent of n, the number of steps in the walk.

It is straightforward to find the mean and variance of X_n from (3.1) with $X_0 = 0$:

$$\mathbf{E}(X_n) = \mathbf{E}\left(\sum_{i=1}^{n} W_i\right) = \sum_{i=1}^{n} \mathbf{E}(W_i),$$

$$\mathbf{V}(X_n) = \mathbf{V}\left(\sum_{i=1}^{n} W_i\right) = \sum_{i=1}^{n} \mathbf{V}(W_i),$$

since the W_i are independent and identically distributed random variables. Thus

$$\mathbf{E}(W_i) = 1.p + (-1).q = p - q.$$

Since

$$\mathbf{V}(W_i) = \mathbf{E}(W_i^2) - [\mathbf{E}(W_i)]^2,$$

and

$$\mathbf{E}(W_i^2) = 1^2.p + (-1)^2 q = p + q = 1,$$

then

$$\mathbf{V}(W_i) = 1 - (p - q)^2 = 4pq.$$

Hence the probability distribution of the position of the random walk at stage n has mean and variance

$$\mathbf{E}(X_n) = n(p - q), \qquad \text{and} \qquad \mathbf{V}(X_n) = 4npq.$$

If $p > \frac{1}{2}$ then we would correctly expect a drift away from the origin in a positive direction, and if $p < \frac{1}{2}$, it would be expected that the drift would be in the negative direction. However, since $\mathbf{V}(X_n)$ is proportional to n, and thus grows with increasing n, we would be increasingly uncertain about the position of the walker as n increases.

For the **symmetric random walk**, when $p = \frac{1}{2}$, the expected position after n steps is the origin. However, this is precisely the value of p which yields the maximum value of the variance $\mathbf{V}(X_n) = 4npq = 4np(1-p)$ (check where $d\mathbf{V}(X_n)/dp = 0$). Thus the maximum value as a function of p is $\max_p \mathbf{V}(X_n) = n$.

Knowing the mean and standard variation of a random variable does not enable us to identify the probability distribution. However, for large n we may apply the central limit theorem, which states: if $W_1, W_2, \ldots$ is a sequence of independent identically distributed (iid) random variables with mean μ and variance σ^2, then, for the random variable $X_n = W_1 + W_2 + \cdots + W_n$,

$$\frac{X_n - n\mu}{\sqrt{n\sigma^2}}$$

has a standard normal $N(0, 1)$ distribution as $n \to \infty$ (see Section 1.8). We shall not give a proof of this theorem in this book (see, for example, Larsen and Marx (1985)). In our case $\mu = p - q$ and $\sigma^2 = \mathbf{V}(W_i) = 4pq$. Put another way, we can say that

$$Z_n = \frac{X_n - n(p - q)}{\sqrt{4npq}} \approx N(0, 1) \text{ or } X_n \sim N[n(p - q), 4npq] \qquad (3.2)$$

for large n.

In this approximation, X_n is a *discrete* random variable, but the normal distribution assumes a continuous random variable. We can overcome this by employing a **continuity correction**. Suppose that we require the probability that the position of the walk at (say) the 100th step with $p = 0.7$ (say) lies on or between positions 35 and 45. Then

$$\mathbf{E}(X_{100}) = 100(0.7 - 0.3) = 40, \quad \mathbf{V}(X_{100}) = 4 \times 100 \times 0.7 \times 0.3 = 84.$$

For the correction we use $\mathbf{P}(34.5 < X_{100} < 45.5)$ since the event $(35 < X_{100} < 45)$ is approximated by $\mathbf{P}(34.5 < X_{100} < 45.5)$ for large n due to rounding. Put another way the event $(35 < X_{100} < 45)$ for x_n discrete is equivalent to $\mathbf{P}(34.5 < X_{100} < 45.5)$ for X_n continuous.

From (3.2)

$$-0.60 \approx \frac{34.5 - 40}{\sqrt{84}} < Z_{100} = \frac{X_{100} - n(p - q)}{\sqrt{4npq}} < \frac{45.5 - 40}{\sqrt{84}} \approx 0.60.$$

Finally

$$\mathbf{P}(-0.60 < Z_{100} < 0.60) = \Phi(0.60) - \Phi(-0.60) = 0.45, \qquad (3.3)$$

where $\Phi(s)$ is the standard normal distribution function. Hence the required probability is approximately 0.45.

3.3 The exact probability distribution of a random walk

As before we assume that the walk is such that $X_0 = 0$, with steps to the right or left occurring with probabilities p and $q = 1 - p$, respectively. The probability distribution of the random variable X_n, the position after n steps, is a more difficult problem. The position X_n, after n steps, can be written as

$$X_n = R_n - L_n,$$

where R_n is the random variable of the number of right (positive) steps $(+1)$ and L_n is that of the number of left (negative) steps (-1). Furthermore,

$$N = R_n + L_n,$$

where N is the random variable of the number of steps. Hence,

$$R_n = \frac{1}{2}(N + X_n).$$

Now, let $v_{n,x}$ be the probability that the walk is at position x after n steps. Thus

$$v_{n,x} = \mathbf{P}(X_n = x). \tag{3.4}$$

The type of distribution can be deduced by the following combinatorial argument. To reach position x after $n \geq |x|$ steps requires $r = \frac{1}{2}(n + x)$ $(+1)$ steps (and consequently $l = n - r = \frac{1}{2}(n - x)$ (-1) steps). Right (r) and left (l) must be integers so that it is implicit that if x is an odd (even) integer then n must also be odd (even). We now ask: in how many ways can $r = \frac{1}{2}(n + x)$ steps be chosen from n? The answer is

$$h_{n,x} = \frac{n!}{r!l!} = \frac{n!}{r!(n-r)!} = \binom{n}{r}.$$

The $r = \frac{1}{2}(n + x)$ steps occur with probability p^r and the $l = \frac{1}{2}(n - x)$ steps with probability q^l. Hence, the probability that the walk is at position x after n steps is (Eqn (3.4))

$$v_{n,x} = h_{n,x}p^r q^l = \binom{n}{r}p^r q^l = \binom{n}{\frac{1}{2}(n+x)}p^{\frac{1}{2}(n+x)}q^{\frac{1}{2}(n-x)}. \tag{3.5}$$

From Section 1.7 we observe that (3.5) defines a *binomial distribution* with index n and probability p.

Example 3.1 *Find the probability that a random walk of 8 steps with probability $p = 0.6$ ends at (a) position $x = 6$, (b) position $x = -4$.*

(a) The events $X_8 = 6$ occur with $r = 7$ positive $(+1)$ steps and $l = 1$ negative (-1) steps but they could be in any order. Hence by (3.5),

$$\mathbf{P}(X_8 = 6) = \binom{8}{7}0.6^7 \times 0.4 = 0.0896.$$

(b) For $X_8 = -4$,

$$\mathbf{P}(X_8 = -4) = \binom{8}{2}0.6^2 \times 0.4^6 = 0.0413.$$

We defined $v_{n,x}$ to be the probability that the walk ends at position x after n steps: the walk could have overshot x before returning there. A related probability is the probability that the *first* visit to position x occurs at the n-th step. This is sometimes also known as the **first passage** through x, and will be considered in the next section for $x = 0$. The following is a descriptive derivation of the associated probability generating function for the symmetric random walk in which the walk starts at the origin, and we consider the probability that the walk is at the origin at a future step.

The walk can only return to the origin if n is even. For this reason put $n = 2m$, ($m = 1, 2, 3, \ldots$). From the previous section a symmetric random walk ($p = \frac{1}{2}$) is at the origin at step $2m$ if (Eqn (3.5))

$$v_{2m,0} = \frac{1}{2^{2m}} \binom{2m}{m} = p_{2m}, \quad (m = 1, 2, 3, \ldots), \tag{3.6}$$

say. Construct a generating function (see Section 1.9) with these coefficients, namely

$$H(s) = \sum_{m=1}^{\infty} p_{2m} s^{2m} = \sum_{m=1}^{\infty} \frac{1}{2^{2m}} \binom{2m}{m} s^{2m}.$$

An alternative identity for the binomial $\binom{2m}{m}$ is needed as follows:

$$
\begin{aligned}
\binom{2m}{m} &= \frac{(2m)!}{m!m!} = \frac{2m(2m-1)(2m-2)\ldots 3.2.1}{m!m!}, \\
&= \frac{2^m m!(2m-1)(2m-3)\ldots 3.1}{m!m!}, \\
&= \frac{2^{2m}}{m!} \frac{1}{2} \frac{3}{2} \left(m - \frac{1}{2}\right)
\end{aligned}
$$

Hence

$$H(s) = \sum_{m=1}^{\infty} \frac{1}{m!} \left[\frac{1}{2} \frac{3}{2} \cdots \left(m - \frac{1}{2}\right)\right] s^{2m}, \tag{3.7}$$

which is recognizable as the binomial expansion of

$$H(s) = (1 - s^2)^{-\frac{1}{2}} - 1.$$

It is evident that $H(1) = \infty$: in other words the series for $H(s)$ diverges at $s = 1$. This breaks the first condition (a) for a generating function in Section 1.9. However, the coefficients in the series do give the correct probabilities. This type of generating function is sometimes known as giving a **defective distribution**. The reason for the defect is that the *mean* [1]

$$H'(1) = \lim_{s \to 1-} = \lim_{s \to 1-} \frac{s}{(1 - s^2)^{\frac{3}{2}}} = \infty. \tag{3.8}$$

This means that the mean number of visits to the origin is infinite, but we shall not

[1] $s \to 1-$ means that s tends to 1 from the left.

prove these conclusions here; see Problem 3.24 for further discussion of the divergence of the series.

3.4 First returns of the symmetric random walk

As we remarked in the previous section, a related probability is that for the event that the *first* visit to position x occurs at the n-th step given that the walk starts at the origin, known also as the **first passage** through x. We shall look in detail at the case $x = 0$, which will lead to the probability of the first return to the origin. We shall approach the first passage by using total probability (Section 1.3).

As in the previous section, for a return to the origin $x = 0$ to exist, the number of steps n must be even, so let $n = 2m$ ($m = 1, 2, 3, \ldots$). Let A_{2m} be the event that the random number $X_{2m} = 0$, and let B_{2k} be the event that the *first return* to the origin is at the $2k$-th step. The significant difference is that the event A_{2m} can occur many times in a given walk. By the law of total probability (Section 1.3),

$$\mathbf{P}(A_{2m}) = \sum_{k=1}^{m} \mathbf{P}(A_{2m}|B_{2k})\mathbf{P}(B_{2k}). \qquad (3.9)$$

(In the earlier notations, $\mathbf{P}(A_{2m}) = v_{2m,0} = p_{2m}$.) What (3.9) states is that the probability that the walk is at the origin at step $2k$ must include *every* previous first return for every k from $k = 1$ to $k = m$. As a consequence, $\mathbf{P}(A_{2m}|B_{2k}) = p_{2m-2k}$ (we define $p_0 = 1$). Let $f_{2k} = \mathbf{P}(B_{2k})$. Our aim is to construct f_{2k} from (3.9).

It is worth pausing to look at (3.8) in a particular simple case to interpret the various probabilities. Suppose $m = 2$. Then there are $2^4 = 16$ possible distinct walks which start at $x = 0$, of which 6 reach $x = 0$ after 4 steps, namely:

- (i) $0 \to 1 \to 0 \to 1 \to 0$;
- (ii) $0 \to 1 \to 0 \to -1 \to 0$;
- (iii) $0 \to -1 \to 0 \to -1 \to 0$;
- (iv) $0 \to -1 \to 0 \to 1 \to 0$;
- (v) $0 \to 1 \to 2 \to 1 \to 0$;
- (vi) $0 \to -1 \to -2 \to -1 \to 0$.

Therefore,

$$f_2 = \mathbf{P}(B_2) = \frac{2}{2^2} = \frac{1}{2} \qquad [0 \to 1 \to 0 \text{ and } 0 \to -1 \to 0],$$

$$f_4 = \mathbf{P}(B_4) = \frac{2}{2^4} = \frac{1}{8} \qquad [\text{from (v)and (vi)}],$$

$$p_{4-2} = p_2 = \mathbf{P}(A_4|B_2) = \frac{2 \times 2^2}{2^4} = \frac{1}{2},$$

$$p_{4-4} = p_0 = \mathbf{P}(A_4|B_4) = \frac{2}{2} = 1 \qquad [\text{certainty}].$$

Hence,

$$\mathbf{P}(A_4) = \tfrac{1}{2}\tfrac{1}{2} + 1\cdot\tfrac{1}{8} = \tfrac{3}{8},$$

which agrees with $v_{4,0}$ in (3.5). The first return probabilities are

$$f_2 = \mathbf{P}(B_2), \qquad f_4 = \mathbf{P}(B_4) = \tfrac{1}{8}.$$

Equation (3.8) can be expressed in the form

$$p_{2m} = \sum_{k=1}^{m} p_{2m-2k} f_{2k} \tag{3.10}$$

with $p_0 = 1$ and $f_0 = 0$. We intend to construct a generating function for the first return or passage probabilities . Multiply both sides of (3.10) by s^{2m} and sum for all $m \geq 1$ (assuming convergence of the infinite series for the moment). From Section 3.3, Eqn (3.7),

$$H(s) - 1 = \sum_{m=1}^{\infty} p_{2m} s^{2m} = \sum_{m=1}^{\infty} \sum_{k=1}^{m} p_{2m-2k} f_{2k} s^{2k}. \tag{3.11}$$

Expanded, the series looks like this (remember, $p_0 = 1$):

$$\begin{aligned}
H(s) &= \sum_{k=1}^{1} p_{2-2k} f_{2k} s^{2k} + \sum_{k=1}^{2} p_{4-2k} f_{2k} s^{2k} + \sum_{k=1}^{3} p_{6-2k} f_{2k} s^{2k} + \cdots \\
&= f_2 s^2 + [p_2 f_2 s^2 + f_4 s^4] + [p_4 f_2 s^2 + p_2 f_4 s^4 + f_6 s^6] + \cdots \\
&= [1 + p_2 s^2 + p_4 s^4 + \cdots][f_2 s^2 + f_4 s^4 + \cdots] \\
&= \left[\sum_{m=0}^{\infty} p_{2m} s^{2m} \right] \left[\sum_{k=1}^{\infty} f_{2k} s^{2k} \right] = [1 + H(s)] F(s), \tag{3.12}
\end{aligned}$$

using (3.10), where

$$F(s) = \sum_{k=1}^{\infty} f_{2k} s^{2k}.$$

The derivation of (3.12) from (3.11) in this descriptive approach is really a well-known result from the formula for the product of two power series. Finally from (3.12) it follows that

$$F(s) = H(s)/[H(s) + 1] = 1 - (1 - s^2)^{\frac{1}{2}}. \tag{3.13}$$

The probability that the walk will, at some step, return to the origin is the sum of all the first returns, namely

$$\sum_{k=1}^{\infty} f_{2k} = F(1) = 1;$$

in other words, return is certain. In this walk the origin (or any starting point by translation) is said to be **persistent**. However, the mean number of steps until this return occurs is

$$\sum_{n=1}^{\infty} n f_n = \lim_{s \to 1-} F'(s) = \lim_{s \to 1-} \frac{s}{(1 - s^2)^{\frac{1}{2}}} = \infty.$$

In other words a symmetric random walk which starts at the origin is certain to return
there in the future, but, on average, it will take an infinite number of steps.

Example 3.2 *Find the probability that a symmetric random walk starting from the origin
returns there for the first time after 6 steps.*

We require the coefficient of s^6 in the power series expansion of the pgf $F(s)$, which is

$$
\begin{aligned}
F(s) &= 1 - (1 - s^2)^{\frac{1}{2}} = 1 - [1 - \frac{1}{2}s^2 - \frac{1}{8}s^4 - \frac{1}{16}s^6 + O(s^8)] \\
&= \frac{1}{2}s^2 + \frac{1}{8}s^4 + \frac{1}{16}s^6 + O(s^8).
\end{aligned}
$$

Hence the probability of a first return at step 6 is $\frac{1}{16}$.

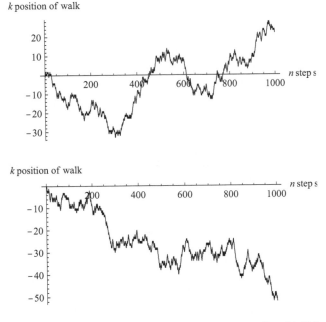

Figure 3.2 *Two computer simulations of symmetric random walks of 1,000 steps each.*

Figure 3.2 shows a computer simulation of two sample walks of 1,000 steps starting at $k = 0$, with forward or backward steps equally likely. It might be expected
intuitively that the walk would tend to oscillate about the starting position $k = 0$
by some law of averages. But a feature of such walks is how few times the walk
recrosses the axis $k = 0$. In fact in the first case, after a brief oscillation about $k = 0$,
the walk does not return to the start and finishes some 70 paces away. Intuition can
be misleading in these problems. Remember that the expected state is the average of
many walks. A full discussion of this phenomena can be found in Feller (1968).

Example 3.3 *A symmetric random walk starts at $x = 0$. Find the probabilities that the walk:*

(a) is at $x = 0$ after 10 steps;
(b) first returns to $x = 0$ after 10 steps;
(c) returns to $x = 0$ on the second occasion after a further 10 steps.

(a) From (3.5) the probability that the walk is at $x = 0$ after 10 steps is

$$v_{10,0} = \frac{1}{2^{10}} \binom{10}{5} = \frac{1}{2^{10}} \frac{10!}{5!5!} = \frac{63}{256} = 0,246\ldots.$$

(b) We require the coefficient of s^{10} in the expansion of $F(s)$ in (3.12). The series is

$$F(s) = \tfrac{1}{2}s^2 + \tfrac{1}{8}s^4 + \tfrac{1}{16}s^6 + \tfrac{5}{128}s^8 + \tfrac{7}{256}s^{10} + \cdots.$$

Hence the probability of a first return at the 10-th step is $f_{10} = 7/256 = 0.027\ldots$
(c) To return to $x = 0$ for a second occasion after 10 steps has the probability

$$q_{10} = f_2 f_8 + f_4 f_6 + f_6 f_4 + f_8 f_2;$$

in other words q_{10} is the probability that a first return occurs at step 2 followed by a subsequent first return occurring at step 8, plus the probability that a first return occurs at step 4 followed by a subsequent first return occurring at step 6, and so on. The result is

$$q_{10} = \tfrac{1}{2}\cdot\tfrac{5}{128} + \tfrac{1}{8}\cdot\tfrac{1}{16} + \tfrac{1}{16}\cdot\tfrac{1}{8} + \tfrac{1}{256}\cdot\tfrac{1}{2} = \tfrac{25}{512} = 0.049\cdots.$$

3.5 Problems

3.1. In a random walk the probability that the walk advances by one step is p and retreats by one step is $q = 1 - p$. At step n let the position of the walker be the random variable X_n. If the walk starts at $x = 0$, enumerate all possible sample paths which lead to the value $X_4 = 2$. Verify that

$$\mathbf{P}[X_4 = -2] = \binom{4}{1}pq^3.$$

3.2. A symmetric random walk starts from the origin. Find the probability that the walker is at the origin at step 8. What is the probability, also at step 8, that the walker is at the origin but that it is not the first visit there?

3.3. An asymmetric walk starts at the origin. From Eqn (3.5), the probability that the walk reaches x in n steps is given by

$$v_{n,x} = \binom{n}{\frac{1}{2}(n+x)} p^{\frac{1}{2}(n+x)} q^{\frac{1}{2}(n-x)},$$

where n and x are both even or both odd. If $n = 4$, show that the mean value of position x is $4(p - q)$, confirming the result in Section 3.2.

3.4. The pgf for the first return distribution $\{f_n\}$ to the origin in a symmetric random walk is given by

$$F(s) = 1 - (1 - s^2)^{\frac{1}{2}}$$

(see Section 3.4).

(a) Using the binomial theorem, find a formula for f_n, the probability that the first return occurs at the n-th step.

(b) What is the variance of the probability distribution f_n, $n = 1, 2, \ldots$?

3.5. An unbiased coin is spun $2n$ times and the sequence of heads and tails is recorded. What is the probability that the number of heads equals the number of tails after $2n$ spins?

3.6. For an asymmetric walk with parameters p and $q = 1 - p$, the probability that the walk is at the origin after n steps is

$$q_{2n} = v_{2n,0} = \binom{2n}{n} p^n q^n, \qquad (n = 1, 2, 3, \ldots),$$

from Eqn (3.5). Note that $\{q_{2n}\}$ is not a probability distribution. Find the mean number of steps of a return to the origin conditional on a return occurring.

3.7. Using Eqn (3.13) relating to the generating functions of the returns and first returns to the origin, namely

$$H(s) = [H(s) + 1]F(s),$$

which is still valid for the asymmetric walk, show that

$$F(s) = 1 - (1 - 4pqs^2)^{\frac{1}{2}},$$

where $p \neq q$. Show that a first return to the origin is not certain unlike the situation in the symmetric walk. Find the mean number of steps to the first return.

3.8. A symmetric random walk starts from the origin. Show that the walk does not revisit the origin in the first $2n$ steps with probability

$$h_n = 1 - f_2 - f_4 - \cdots - f_{2n},$$

where f_{2n} is the probability that a first return occurs at the $2n$-th step.

The generating function for the sequence $\{f_n\}$ is

$$F(s) = 1 - (1 - s^2)^{\frac{1}{2}}$$

(see Section 3.4). Show that

$$f_2 = \frac{1}{2} \qquad f_{2n} = \frac{(2n-3)!}{n!(n-2)!2^{2n-2}}, \qquad (n = 2, 3, \ldots).$$

Show that h_n satisfies the first-order difference equation

$$h_{n+1} - h_n = f_{2n+2} = \frac{(2n+1)!}{(n+1)!(n-1)!2^{2n}}.$$

Verify that this equation has the general solution

$$h_n = C + \binom{2n}{n} \frac{1}{2^{2n}},$$

where C is a constant. By calculating h_1, confirm that the probability of no return to the origin in the first $2n$ steps is

$$\binom{2n}{n} \frac{1}{2^{2n}}.$$

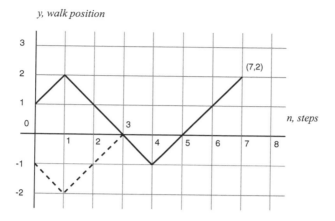

Figure 3.3 *Representation of a random walk.*

3.9. A walk can be represented as a connected graph between coordinates (n, y) where the ordinate y is the position on the walk, and the abscissa n represents the number of steps. A walk of 7 steps which joins $(0, 1)$ and $(7, 2)$ is shown in Figure 3.3. Suppose that a walk starts at $(0, y_1)$ and finishes at (n, y_2), where $y_1 > 0$, $y_2 > 0$, and $n + y_2 - y_1$ is an even number. Suppose also that the walk first visits the origin at $n = n_1$. Reflect that part of the path for which $n \leq n_1$ in the n-axis (see Figure 3.3), and use a reflection argument to show that the number of paths from $(0, y_1)$ to (n, y_2) which touch or cross the n-axis equals the number of *all* paths from $(0, -y_1)$ to (n, y_2). This is known as the **reflection principle**.

3.10. A walk starts at $(0, 1)$ and returns to $(2n, 1)$ after $2n$ steps. Using the reflection principle (see Problem 3.9), show that there are

$$\frac{(2n)!}{n!(n+1)!}$$

different paths between the two points which do not ever revisit the origin. What is the probability that the walk ends at $(2n, 1)$ after $2n$ steps without ever visiting the origin, assuming that the random walk is symmetric?

Show that the probability that the first visit to the origin after $2n + 1$ steps is

$$p_n = \frac{1}{2^{2n+1}} \frac{(2n)!}{n!(n+1)!}.$$

3.11. A symmetric random walk starts at the origin. Let $f_{n,1}$ be the probability that the first visit to position $x = 1$ occurs at the n-th step. Obviously, $f_{2n,1} = 0$. The result from Problem 3.10 can be adapted to give

$$f_{2n+1,1} = \frac{1}{2^{2n+1}} \frac{(2n)!}{n!(n+1)!}, \qquad (n = 0, 1, 2, \ldots).$$

Suppose that its pgf is

$$G_1(s) = \sum_{n=0}^{\infty} f_{2n+1,1} s^{2n+1}.$$

Show that

$$G_1(s) = [1 - (1 - s^2)^{\frac{1}{2}}]/s.$$

[Hint: the identity

$$\frac{1}{2^{2n+1}} \frac{(2n)!}{n!(n+1)!} = (-1)^n \binom{\frac{1}{2}}{n+1}, \qquad (n = 0, 1, 2, \ldots)$$

is useful in the derivation of $G_1(s)$.]

Show that any walk starting at the origin is certain to visit $x > 0$ at some future step, but that the mean number of steps in achieving this is infinite.

3.12. A symmetric random walk starts at the origin. Let $f_{n,x}$ be the probability that the first visit to position x occurs at the n-th step (as usual, $f_{n,x} = 0$ if $n + x$ is an odd number). Explain why

$$f_{n,x} = \sum_{k=1}^{n-1} f_{n-k,x-1} f_{k,1}, \qquad (n \geq x > 1).$$

If $G_x(s)$ is its pgf, deduce that

$$G_x(s) = \{G_1(s)\}^x,$$

where $G_1(s)$ is given explicitly in Problem 3.11. What are the probabilities that the walk first visits $x = 3$ at the steps $n = 3$, $n = 5$, and $n = 7$?

3.13. Problem 3.12 looks at the probability of a first visit to position $x \geq 1$ at the n-th step in a symmetric random walk which starts at the origin. Why is the pgf for the first visit to position x where $|x| \geq 1$ given by

$$G_x(s) = \{G_1(s)\}^{|x|},$$

where $G_1(s)$ is defined in Problem 3.11?

3.14. An asymmetric walk has parameters p and $q = 1 - p \neq p$. Let $g_{n,1}$ be the probability that the first visit to $x = 1$ occurs at the n-th step. As in Problem 3.11, $g_{2n,1} = 0$. It was effectively shown in Problem 3.10 that the number of paths from the origin, which return to the origin after $2n$ steps, is

$$\frac{(2n)!}{n!(n+1)!}.$$

Explain why

$$g_{2n+1,1} = \frac{(2n)!}{n!(n+1)!} p^{n+1} q^n.$$

Suppose that its pgf is

$$G_1(s) = \sum_{n=0}^{\infty} g_{2n+1,1} s^{2n+1}.$$

Show that

$$G_1(s) = [1 - (1 - 4pqs^2)^{\frac{1}{2}}]/(2qs).$$

(The identity in Problem 3.11 is required again.)

What is the probability that the walk ever visits $x > 0$? How does this result compare with that for the symmetric random walk?

What is the pgf for the distribution of first visits of the walk to $x = -1$ at step $2n + 1$?

3.15. It was shown in Section 3.3 that, in a random walk with parameters p and $q = 1 - p$, the probability that a walk is at position x at step n is given by

$$v_{n,x} = \binom{n}{\frac{1}{2}(n + x)} p^{\frac{1}{2}(n+x)} q^{\frac{1}{2}(n-x)}, \qquad |x| \leq n,$$

where $\frac{1}{2}(n + x)$ must be an integer. *Verify* that $v_{n,x}$ satisfies the difference equation

$$v_{n+1,x} = p v_{n,x-1} + q v_{n,x+1},$$

subject to the initial conditions

$$v_{0,0} = 1, \qquad v_{n,x} = 0, \quad (x \neq 0).$$

Note that this difference equation has differences of two arguments.

Can you develop a direct argument which justifies the difference equation for the random walk?

3.16. In the usual notation, $v_{2n,0}$ is the probability that, in a symmetric random walk, the walk visits the origin after $2n$ steps. Using the difference equation from Problem 3.15, $v_{2n,0}$ satisfies

$$v_{2n,0} = \tfrac{1}{2} v_{2n-1,-1} + \tfrac{1}{2} v_{2n-1,1} = v_{2n-1,1}.$$

How can the last step be justified? Let

$$G_1(s) = \sum_{n=1}^{\infty} v_{2n-1,1} s^{2n-1}$$

be the pgf of the distribution $\{v_{2n-1,1}\}$. Show that

$$G_1(s) = [(1 - s^2)^{-\frac{1}{2}} - 1]/s.$$

By expanding $G_1(s)$ as a power series in s show that

$$v_{2n-1,1} = \binom{2n-1}{n} \frac{1}{2^{2n-1}}.$$

By a repetition of the argument show that

$$G_2(s) = \sum_{n=0}^{\infty} v_{2n,2} s^{2n} = [(2 - s^2)(1 - s^2)^{-\frac{1}{2}} - 2]/s^2.$$

3.17. A random walk takes place on a circle which is marked out with n positions. Thus, as shown in Figure 3.4, position n is the same as position 0. This is known as the **cyclic random walk** of **period** n. A symmetric random walk starts at 0. What is the probability that the walk is at 0 after j steps in the cases:
(a) $j < n$;
(b) $n \leq j < 2n$?
Distinguish carefully the cases in which j and n are even or odd.

3.18. An unrestricted random walk with parameters p and q starts from the origin, and lasts for 50 paces. Estimate the probability that the walk ends at 12 or more paces from the origin in the cases:
(a) $p = q = \frac{1}{2}$;
(b) $p = 0.6$, $q = 0.4$.

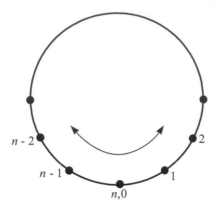

Figure 3.4 *The cyclic random walk of period n.*

3.19. In an unrestricted random walk with parameters p and q, for what value of p are the mean and variance of the probability distribution of the position of the walk at stage n the same?

3.20. Two walkers each perform symmetric random walks with synchronized steps, both starting from the origin at the same time. What is the probability that they are both at the origin at step n?

3.21. A random walk takes place on a two-dimensional lattice as shown in Figure 3.5. In the example shown the walk starts at $(0, 0)$ and ends at $(2, -1)$ after 13 steps. In this walk di-

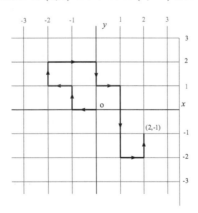

Figure 3.5 *A two-dimensional random walk.*

rect diagonal steps are not permitted. We are interested in the probability that the symmetric random walk, which starts at the origin, has returned there after $2n$ steps. Symmetry in the two-dimensional walk means that there is a probability of $\frac{1}{4}$ that, at any position, the walk goes right, left, up, or down at the next step. The total number of different walks of length $2n$ which start at the origin is 4^{2n}. For the walk considered, the number of right steps (positive x direction) must equal the number of left steps, and the number of steps up (positive y direction) must equal those down. Also the number of right steps must range from 0 to n, and the corresponding steps up must range from n to 0. Explain why the probability that the walk returns to the origin after $2n$ steps is

$$p_{2n} = \frac{(2n)!}{4^{2n}} \sum_{r=0}^{n} \frac{1}{[r!(n-r)!]^2}.$$

Prove the two identities

$$\frac{(2n)!}{[r!(n-r)!]^2} = \binom{2n}{n}\binom{n}{r}^2, \qquad \binom{2n}{n} = \sum_{r=0}^{n} \binom{n}{r}^2.$$

[Hint: compare the coefficients of x^n in $(1+x)^{2n}$ and $[(1+x)^n]^2$.] Hence show that

$$p_{2n} = \frac{1}{4^{2n}} \binom{2n}{n}^2.$$

Calculate $1/(\pi p_{40})$ and $1/(\pi p_{80})$. How would you guess that p_{2n} behaves for large n?

3.22. A random walk takes place on the positions $\{\ldots, -2, -1, 0, 1, 2, \ldots\}$. The walk starts at 0. At step n, the walker has a probability q_n of advancing one position, or a probability $1 - q_n$ of retreating one step (note that the probability depends on the step not the position of the walker). Find the expected position of the walker at step n. Show that if $q_n = \frac{1}{2} + r_n$, $(-\frac{1}{2} < r_n < \frac{1}{2})$, and the series $\sum_{j=1}^{\infty} r_j$ is convergent, then the expected position of the walk will remain finite as $n \to \infty$.

3.23. A symmetric random walk starts at k on a position chosen from $0, 1, 2, \ldots, a$, where $0 < k < a$. As in the gambler's ruin problem, the walk stops whenever 0 or a is first reached. Show that the expected number of visits to position j where $0 < j < k$ is $2j(a-k)/a$ before the walk stops.

CHAPTER 4

Markov Chains

4.1 States and transitions

The random walk discussed in the previous chapter is a special case of a more general **Markov process**[1]. Suppose that a random process passes through a discrete sequence of **steps** or **trials** numbered $n = 0, 1, 2, \ldots$, where the outcome of the n-th trial is the random variable X_n; X_0 is the initial position of the process. This discrete random variable can take one of the values $i = 1, 2, \ldots m$. The actual outcomes are called the **states** of the system, and are denoted by E_i ($i = 1, 2, \ldots m$) (states can be any consecutive sequence of integers, say starting with $i = 0$ or some other integer, but they can be renumbered to start with $i = 1$). In most but not quite all cases in this text, we shall investigate systems with a *finite* number, m, of states $E_1, E_2, \ldots E_m$, which are independent and exhaustive.

If the random variables $X_{n-1} = i$ and $X_n = j$, then the system has made a **transition** $E_i \rightarrow E_j$, that is, a transition from state E_i to state E_j at the n-th trial. Note that i can equal j, so that transitions within the same state may be possible. We need to assign probabilities to the transitions $E_i \rightarrow E_j$. This discrete process is known as a **chain**. Generally in chains the probability that $X_n = j$ will depend on the whole sequence of random variables starting with the initial value X_0. The **Markov chain** has the characteristic property that the probability that $X_n = j$ depends *only* on the immediate previous state of the system. Formally this means that we need no further information at each step other than, for each i and j,

$$\mathbf{P}\{X_n = j | X_{n-1} = i\},$$

which means the probability that $X_n = j$ given that $X_{n-1} = i$: this probability is *independent* of the values of $X_{n-2}, X_{n-3}, \ldots, X_1$. Put alternatively, conditional on the present state of the chain, its future and present are independent.

In some chains the probabilities $\mathbf{P}\{X_n = j | X_{n-1} = i\}$ are functions of n, the step or trial number. If this is not the case, so that the probabilities are the same at every step, then the chain is said to be **homogeneous**.

---

[1] Andrey Markov (1856–1922), Russian mathematician.

4.2 Transition probabilities

For a finite Markov chain with m states $E_1, E_2, \ldots, E_m$, introduce the notation

$$p_{ij} = \mathbf{P}\{X_n = j | X_{n-1} = i\}, \tag{4.1}$$

where $i, j = 1, 2, \ldots, m$ to represent the probability of a transition from state E_i to E_j. The numbers p_{ij} are known as the **transition probabilities** of the chain, and must satisfy

$$p_{ij} \geq 0, \quad \sum_{j=1}^{m} p_{ij} = 1$$

for each $i = 1, 2, \ldots, m$. If $p_{ij} > 0$, then we say that state E_i can **communicate** with E_j: two-way communication is possible if additionally $p_{ji} > 0$ (see Section 4.7(a) later). Obviously for each fixed i, the list $\{p_{ij}\}$ is a **probability distribution**, since at any step one of the outcomes $E_1, E_2, \ldots, E_m$ must occur: the states E_i, $(i = 1, 2, \ldots m)$.

Transition probabilities form an $m \times m$ array which can be assembled into a **transition matrix** T, where

$$T = [p_{ij}] = \begin{bmatrix} p_{11} & p_{12} & \cdots & p_{1m} \\ p_{21} & p_{22} & \cdots & p_{2m} \\ \vdots & \vdots & \ddots & \vdots \\ p_{m1} & p_{m2} & \cdots & p_{mm} \end{bmatrix}. \tag{4.2}$$

Note that each row of T is a probability distribution. Any square matrix for which $p_{ij} \geq 0$ and $\sum_{j=1}^{m} p_{ij} = 1$ is said to be **row-stochastic**.

Example 4.1. *The matrices $A = [a_{ij}]$ and $B = [b_{ij}]$ are $m \times m$ row-stochastic matrices. Show that $C = AB$ is also row-stochastic.*

By the multiplication rule for matrices

$$C = AB = [a_{ij}][b_{ij}] = \left[\sum_{k=1}^{m} a_{ik} b_{kj} \right].$$

Hence c_{ij}, the general element of C, is given by

$$c_{ij} = \sum_{k=1}^{m} a_{ik} b_{kj}.$$

Since $a_{ij} \geq 0$ and $b_{ij} \geq 0$ for all $i, j = 1, 2, \ldots, m$, it follows that $c_{ij} \geq 0$. Also

$$\sum_{j=1}^{m} c_{ij} = \sum_{j=1}^{m} \sum_{k=1}^{m} a_{ik} b_{kj} = \sum_{k=1}^{m} a_{ik} \sum_{j=1}^{m} b_{kj} = \sum_{k=1}^{m} a_{ik} \cdot 1 = 1,$$

since $\sum_{j=1}^{m} b_{kj} = 1$ and $\sum_{k=1}^{m} a_{ik} = 1$.

It follows from this example that any power T^n of the transition matrix T must also be row-stochastic.

(i) **The absolute probability** $p_j^{(n)}$

One further probability which will be of interest is the probability of outcome E_j after n steps, given an **initial probability distribution** $\{p_i^{(0)}\}$. Here $p_i^{(0)}$ is the probability that initially the system occupies state E_i. Of course we must have $\sum_{i=1}^m p_i^{(0)} = 1$. Let $p_j^{(1)}$ be the probability E_j is occupied after one step. Then, by the law of total probability (see Section 1.3)

$$p_j^{(1)} = \sum_{i=1}^m p_i^{(0)} p_{ij}. \tag{4.3}$$

We can express this more conveniently in vector form. Let $\mathbf{p}^{(0)}$ and $\mathbf{p}^{(1)}$ be the **probability (row) vectors** given by

$$\mathbf{p}^{(0)} = \begin{bmatrix} p_1^{(0)} & p_2^{(0)} & \cdots & p_m^{(0)} \end{bmatrix} \tag{4.4}$$

and

$$\mathbf{p}^{(1)} = \begin{bmatrix} p_1^{(1)} & p_2^{(1)} & \cdots & p_m^{(1)} \end{bmatrix}. \tag{4.5}$$

Here $\mathbf{p}^{(0)}$ is the initial distribution, and the components of $\mathbf{p}^{(1)}$ will be the probabilities that each of the states $E_1, E_2, \ldots, E_m$ is reached after one step. Equation (4.3) can be represented as a matrix product as follows:

$$\mathbf{p}^{(1)} = \left| p_j^{(1)} \right| = \left[\sum_{i=1}^m p_i^{(0)} p_{ij} \right] = \mathbf{p}^{(0)} T,$$

where T is the transition matrix given by (4.2). If $\mathbf{p}^{(2)}$ is the distribution after two steps, then

$$\mathbf{p}^{(2)} = \mathbf{p}^{(1)} T = \mathbf{p}^{(0)} T T = \mathbf{p}^{(0)} T^2.$$

Hence after n steps by repeating the process

$$\mathbf{p}^{(n)} = \mathbf{p}^{(n-1)} T = \mathbf{p}^{(0)} T^n, \tag{4.6}$$

where

$$\mathbf{p}^{(n)} = \begin{bmatrix} p_1^{(n)} & p_2^{(n)} & \cdots & p_m^{(n)} \end{bmatrix}. \tag{4.7}$$

More generally,

$$\mathbf{p}^{(n+r)} = \mathbf{p}^{(r)} T^n.$$

In (4.7), the component $p_j^{(n)}$ is the **absolute** or **unconditional probability of outcome** E_j at the n-th step given the initial distribution $\mathbf{p}^{(0)}$, that is, $\mathbf{P}\{X_n = j\} = p_j^{(n)}$. Note that

$$\sum_{j=1}^m p_j^{(n)} = 1.$$

Example 4.2. *In a three-state Markov chain with states E_1, E_2, E_3, the chain starts in E_2 so that $\mathbf{p}^{(0)} = \begin{bmatrix} 0 & 1 & 0 \end{bmatrix}$. Find the absolute probability $\mathbf{p}^{(3)}$ if the transition matrix is*

$$T = \begin{bmatrix} \frac{1}{2} & \frac{1}{4} & \frac{1}{4} \\ 0 & \frac{1}{2} & \frac{1}{2} \\ \frac{3}{4} & \frac{1}{4} & 0 \end{bmatrix}.$$

We require

$$T^3 = \begin{bmatrix} \frac{1}{2} & \frac{1}{4} & \frac{1}{4} \\ 0 & \frac{1}{2} & \frac{1}{2} \\ \frac{3}{4} & \frac{1}{4} & 0 \end{bmatrix}^3 = \frac{1}{16} \begin{bmatrix} 7 & 5 & 4 \\ 6 & 6 & 4 \\ 6 & 5 & 5 \end{bmatrix}.$$

Hence,

$$\mathbf{p}^{(3)} = \mathbf{p}^{(0)} T^3 = \frac{1}{16} \begin{bmatrix} 0 & 1 & 0 \end{bmatrix} \begin{bmatrix} 7 & 5 & 4 \\ 6 & 6 & 4 \\ 6 & 5 & 5 \end{bmatrix} = \frac{1}{16} \begin{bmatrix} 6 & 6 & 4 \end{bmatrix}.$$

This result gives the probabilities that, given that the chain starts in E_1, it is in states E_1, E_2, E_3 after 3 steps.

(ii) *The n-step transition probability $p_{ij}^{(n)}$*

We now define $p_{ij}^{(n)}$ as the probability that the chain is in state E_j after n steps *given that the chain started in* state E_i. The first step transition probabilities $p_{ij}^{(1)} = p_{ij}$ are simply the elements of the transition matrix T. We intend to find a formula for $p_{ij}^{(n)}$. Now, by definition,

$$p_{ij}^{(n)} = \mathbf{P}(X_n = j | X_0 = i),$$

and also

$$p_{ij}^{(n)} = \sum_{k=1}^{m} \mathbf{P}(X_n = j, X_{n-1} = k | X_0 = i)$$

for $n \geq 2$, since the chain must have passed through one of all the m possible states at step $n - 1$.

For any three events A, B, and C, we have available the identity

$$\mathbf{P}(A \cap B | C) = \mathbf{P}(A | B \cap C)\mathbf{P}(B | C)$$

(see Example 1.4). Interpreting A as $X_n = j$, B as $X_{n-1} = k$, and C as $X_0 = i$, it follows that

$$\begin{aligned} p_{ij}^{(n)} &= \mathbf{P}(A \cap B | C) = \mathbf{P}(X_n = j, X_{n-1} = k | X_0 = i) \\ &= \sum_{k=1}^{m} \mathbf{P}(X_n = j | X_{n-1} = k, X_0 = i)\mathbf{P}(X_{n-1} = k | X_0 = i) \\ &= \sum_{k=1}^{m} \mathbf{P}(X_n = j | X_{n-1} = k)\mathbf{P}(X_{n-1} = k | X_0 = i) \\ &= \sum_{k=1}^{m} p_{kj}^{(1)} p_{ik}^{(n-1)}, \end{aligned} \tag{4.8}$$

using the Markov property again. These are known as the **Chapman–Kolmogorov equations** [2]. Putting n successively equal to $2, 3, \ldots$, we find that the matrices with

[2] Sydney Chapman (1888–1970), British scientist; Andrey Kolmogorov (1908–1987), Russian mathematician.

these elements are, using the product rule for matrices,

$$\left[p_{ij}^{(2)}\right] = \left[\sum_{k=1}^{m} p_{ik}^{(1)} p_{kj}^{(1)}\right] = T^2,$$

$$\left[p_{ij}^{(3)}\right] = \left[\sum_{k=1}^{m} p_{ik}^{(2)} p_{kj}^{(1)}\right] = T^2 T = T^3,$$

since $p_{ik}^{(2)}$ are the elements of T^2, and so on. Generalising this rule,

$$[p_{ij}^{(n)}] = T^n.$$

Example 4.3. *In a certain region the weather patterns have the following sequence. A day is described as sunny (S) if the sun shines for more than 50% of daylight hours and cloudy (C) if the sun shines for less than 50% of daylight hours. Data indicate that if it is cloudy one day then it is equally likely to be cloudy or sunny on the next day; if it is sunny there is a probability $\frac{1}{3}$ that it is cloudy and $\frac{2}{3}$ that it is sunny the next day.*
(i) Construct the transition matrix T for this process.
(ii) If it is cloudy today, what are the probabilities that it is (a) cloudy, (b) sunny, in three days' time?
(iii) Compute T^5 and T^{10}. How do you think that T^n behaves as $n \to \infty$? How does $\mathbf{p}^{(n)}$ behave as $n \to \infty$? Do you expect the limit to depend on $\mathbf{p}^{(0)}$?

(i) It is assumed that the process is Markov and homogeneous so that transition probabilities depend only on the state of the weather on the previous day. This is a two-state Markov chain with states

$$E_1 = (\text{weather cloudy}, C), \qquad E_2 = (\text{weather sunny}, S).$$

The transition probabilities can be represented by the table below which defines the transition matrix T:

	C	S
C	$\frac{1}{2}$	$\frac{1}{2}$
S	$\frac{1}{3}$	$\frac{2}{3}$

$$\text{or} \quad T = \begin{bmatrix} \frac{1}{2} & \frac{1}{2} \\ \frac{1}{3} & \frac{2}{3} \end{bmatrix}.$$

The actual transition probabilities are

$$p_{11} = \tfrac{1}{2}, \quad p_{12} = \tfrac{1}{2}, \quad p_{21} = \tfrac{1}{3}, \quad p_{22} = \tfrac{2}{3}.$$

(ii) Measuring steps from today, we define

$$\mathbf{p}^{(0)} = \begin{bmatrix} p_1^{(0)} & p_2^{(0)} \end{bmatrix} = \begin{bmatrix} 1 & 0 \end{bmatrix}$$

which means that it is cloudy today. In three days' time,

$$\mathbf{p}^{(3)} = \mathbf{p}^{(0)} T^3 = \begin{bmatrix} 1 & 0 \end{bmatrix} \begin{bmatrix} \frac{1}{2} & \frac{1}{2} \\ \frac{1}{3} & \frac{2}{3} \end{bmatrix}^3$$

$$= \begin{bmatrix} 1 & 0 \end{bmatrix} \begin{bmatrix} 29/72 & 43/72 \\ 43/108 & 65/108 \end{bmatrix}$$

$$= \begin{bmatrix} 29/72 & 43/72 \end{bmatrix} = \begin{bmatrix} 0.403 & 0.600 \end{bmatrix}.$$

Hence the probabilities of cloudy or sunny weather in three days' time are respectively:

(a) $p_1^{(3)} = 29/72$

(b) $p_2^{(3)} = 43/72$.

(iii) The computed values of T^5 and T^{10} are (to 6 decimal places):

$$T^5 = \begin{bmatrix} 0.400077 & 0.599923 \\ 0.399949 & 0.600051 \end{bmatrix}, \qquad T^{10} = \begin{bmatrix} 0.400000 & 0.600000 \\ 0.400000 & 0.600000 \end{bmatrix}.$$

Powers of matrices can be easily computed using software such as R or *Mathematica*. It appears that

$$T^n \to \begin{bmatrix} 0.4 & 0.6 \\ 0.4 & 0.6 \end{bmatrix} = Q,$$

say, as $n \to \infty$. From Eqn (4.6)

$$\mathbf{p}^{(n)} = \mathbf{p}^{(0)} T^n.$$

If $T^n \to Q$ as $n \to \infty$, then we might expect

$$
\begin{aligned}
\mathbf{p}^{(n)} \to \mathbf{p}^{(0)} Q &= \begin{bmatrix} p_1^{(0)} & p_2^{(0)} \end{bmatrix} \begin{bmatrix} 0.4 & 0.6 \\ 0.4 & 0.6 \end{bmatrix} \\
&= \begin{bmatrix} (p_1^{(0)} + p_2^{(0)})0.4 & (p_1^{(0)} + p_2^{(0)})0.6 \end{bmatrix} \\
&= \begin{bmatrix} 0.4 & 0.6 \end{bmatrix}
\end{aligned}
$$

since $p_1^{(0)} + p_2^{(0)} = 1$. Note that $\lim_{n \to \infty} \mathbf{p}^n$ is independent of $\mathbf{p}^{(0)}$. The limit indicates that, in the long run, 40% of days are cloudy and 60% are sunny.

This example indicates that it would be useful if we had a general algebraic formula for the n-th power of a matrix. The algebra required for this aspect of Markov chains will be looked at in the next two sections.

4.3 General two-state Markov chains

Consider the two-state chain with transition matrix

$$T = \begin{bmatrix} 1 - \alpha & \alpha \\ \beta & 1 - \beta \end{bmatrix}, \qquad 0 < \alpha, \beta < 1.$$

We want to construct a formula for T^n. First find the **eigenvalues** (λ) of T: they are given by the solutions of the determinant equation $\det(T - \lambda I_2) = 0$, that is

$$\begin{vmatrix} 1 - \alpha - \lambda & \alpha \\ \beta & 1 - \beta - \lambda \end{vmatrix} = 0, \quad \text{or} \quad (1 - \alpha - \lambda)(1 - \beta - \lambda) - \alpha\beta = 0.$$

Hence λ satisfies the quadratic equation

$$\lambda^2 - \lambda(2 - \alpha - \beta) + 1 - \alpha - \beta = 0,$$

or

$$(\lambda - 1)(\lambda - 1 + \alpha + \beta) = 0. \tag{4.9}$$

Denote the eigenvalues of T by $\lambda_1 = 1$, and $\lambda_2 = 1 - \alpha - \beta = s$, say. *Note that stochastic matrices always have a unit eigenvalue.* We now find the **eigenvectors** associated with each eigenvalue. Let $\mathbf{r}_1$ be the (column) eigenvector of λ_1 which is defined by

$$(T - \lambda_1 I_2)\mathbf{r}_1 = \mathbf{0}, \quad \text{or} \quad \begin{bmatrix} 1 - \alpha - \lambda_1 & \alpha \\ \beta & 1 - \beta - \lambda_1 \end{bmatrix} \mathbf{r}_1 = \mathbf{0},$$

or

$$\begin{bmatrix} -\alpha & \alpha \\ \beta & -\beta \end{bmatrix} \mathbf{r}_1 = \mathbf{0}.$$

Choose *any* (nonzero) solution of this equation, say

$$\mathbf{r}_1 = \begin{bmatrix} 1 \\ 1 \end{bmatrix}.$$

Note that the eigenvector associated with $\lambda_1 = 1$ is always a column of 1's, the result following from the fact that T is stochastic.

Similarly the second eigenvector $\mathbf{r}_2$ satisfies

$$\begin{bmatrix} 1 - \alpha - \lambda_2 & \alpha \\ \beta & 1 - \beta - \lambda_2 \end{bmatrix} \mathbf{r}_2 = \mathbf{0}, \quad \text{or} \quad \begin{bmatrix} \beta & \alpha \\ \beta & \alpha \end{bmatrix} \mathbf{r}_2 = \mathbf{0}.$$

In this case we can choose

$$\mathbf{r}_2 = \begin{bmatrix} -\alpha \\ \beta \end{bmatrix}.$$

Now form the matrix C which has the eigenvectors $\mathbf{r}_1$ and $\mathbf{r}_2$ as columns, so that

$$C = [\ \mathbf{r}_1 \quad \mathbf{r}_2\ ] = \begin{bmatrix} 1 & -\alpha \\ 1 & \beta \end{bmatrix}. \tag{4.10}$$

Now find the inverse matrix C^{-1} of C:

$$C^{-1} = \frac{1}{\alpha + \beta} \begin{bmatrix} \beta & \alpha \\ -1 & 1 \end{bmatrix}.$$

If we now expand the matrix product $C^{-1}TC$, we find that

$$
\begin{aligned}
C^{-1}TC &= \frac{1}{\alpha + \beta} \begin{bmatrix} \beta & \alpha \\ -1 & 1 \end{bmatrix} \begin{bmatrix} 1 - \alpha & \alpha \\ \beta & 1 - \beta \end{bmatrix} \begin{bmatrix} 1 & -\alpha \\ 1 & \beta \end{bmatrix} \\
&= \frac{1}{\alpha + \beta} \begin{bmatrix} \beta & \alpha \\ -1 & 1 \end{bmatrix} \begin{bmatrix} 1 & -\alpha s \\ 1 & \beta s \end{bmatrix} \\
&= \begin{bmatrix} 1 & 0 \\ 0 & s \end{bmatrix} = \begin{bmatrix} \lambda_1 & 0 \\ 0 & \lambda_2 \end{bmatrix} = D, \tag{4.11}
\end{aligned}
$$

say. Now D is a **diagonal matrix** with the eigenvalues of T as its diagonal elements: this process is known in linear algebra as the **diagonalization of a matrix**. The result is significant since diagonal matrices are easy to multiply. From (4.21), if we premultiply D by matrix C and post multiply by C^{-1}, then we find that $T = CDC^{-1}$. Thus

$$
\begin{aligned}
T^2 &= (CDC^{-1})(CDC^{-1}) = (CD)(C^{-1}C)(DC^{-1}) \\
&= (CD)I_2(DC^{-1}) = CDDC^{-1} = CD^2C^{-1},
\end{aligned}
$$

where

$$D^2 = \begin{bmatrix} \lambda_1^2 & 0 \\ 0 & \lambda_2^2 \end{bmatrix} = \begin{bmatrix} 1 & 0 \\ 0 & s^2 \end{bmatrix}.$$

It can be proved by induction that

$$T^n = CD^nC^{-1}, \tag{4.12}$$

where

$$D^n = \begin{bmatrix} \lambda_1^n & 0 \\ 0 & \lambda_2^n \end{bmatrix} = \begin{bmatrix} 1 & 0 \\ 0 & s^n \end{bmatrix}.$$

The product of the matrices in (4.12) can be expanded to give

$$T^n = CD^nC^{-1} = \frac{1}{\alpha + \beta} \begin{bmatrix} 1 & -\alpha \\ 1 & \beta \end{bmatrix} \begin{bmatrix} 1 & 0 \\ 0 & s^n \end{bmatrix} \begin{bmatrix} \beta & \alpha \\ -1 & 1 \end{bmatrix}. \tag{4.13}$$

Since $0 < \alpha, \beta < 1$, it follows that $|s| < 1$, and consequently that $s^n \to 0$ as $n \to \infty$. Hence, from (4.13),

$$T^n \to \frac{1}{\alpha + \beta} \begin{bmatrix} 1 & -\alpha \\ 1 & \beta \end{bmatrix} \begin{bmatrix} 1 & 0 \\ 0 & 0 \end{bmatrix} \begin{bmatrix} \beta & \alpha \\ -1 & 1 \end{bmatrix} = \frac{1}{\alpha + \beta} \begin{bmatrix} \beta & \alpha \\ \beta & \alpha \end{bmatrix}.$$

Further, for *any* initial probability distribution $\mathbf{p}_0$, the distribution over the states after n steps is given by (see Eqn (4.6))

$$\begin{aligned} \mathbf{p}^{(n)} &= \mathbf{p}^{(0)}T^n = \begin{bmatrix} p_1^{(0)} & p_2^{(0)} \end{bmatrix} T^n \\ &\to \begin{bmatrix} p_1^{(0)} & p_2^{(0)} \end{bmatrix} \frac{1}{\alpha + \beta} \begin{bmatrix} \beta & \alpha \\ \beta & \alpha \end{bmatrix} = \begin{bmatrix} \dfrac{\beta}{\alpha + \beta} & \dfrac{\alpha}{\alpha + \beta} \end{bmatrix}, \end{aligned} \tag{4.14}$$

as $n \to \infty$, and the limit is independent of $\mathbf{p}^{(0)}$. The limiting distribution in (4.14) is usually denoted by π. It satisfies

$$\pi = \pi T,$$

which is an example of an **invariant distribution** of the Markov chain, since it is *independent of the initial distribution*. The chain is said to be in **equilibrium**.

If $\alpha = \beta = 1$, then result (4.13) still holds with $s = -1$, but T^n no longer has a limit as $n \to \infty$ but oscillates between two matrices. However, limits for T^n exist if $\alpha = 1, 0 < \beta < 1$ or $0 < \alpha < 1, \beta = 1$.

The conditions for an invariant distribution in the two-state chain raises the question: what are the conditions for the n-state chain? We will return to this later.

4.4 Powers of the general transition matrix

The method derived for the two-state chain in the previous section can be generalized to m-state chains. We shall sketch the diagonalization method here. Let T be an $m \times m$ stochastic matrix, in other words, a possible transition matrix. The eigenvalues of T are given by

$$|T - \lambda I_m| = 0, \tag{4.15}$$

where I_m is the identity matrix of order m. Assume that the eigenvalues are *distinct*, and denoted by $\lambda_1, \lambda_2, \ldots, \lambda_m$. Again note that a stochastic matrix T always has a *unit eigenvalue*, say $\lambda_1 = 1$, with a corresponding unit eigenvector

$$\mathbf{r}_1 = [\ 1 \quad 1 \quad \cdots \quad 1\ ]^t.$$

This follows since every row in $[T - I_m]\mathbf{r}_1$ is zero. [3]
   The corresponding eigenvectors $\mathbf{r}_i$ satisfy the equations

$$[T - \lambda_i I_m]\mathbf{r}_i = \mathbf{0}, \quad (i = 1, 2, \ldots, m). \tag{4.16}$$

Construct the matrix

$$C = [\ \mathbf{r}_1 \quad \mathbf{r}_2 \quad \cdots \quad \mathbf{r}_m\ ],$$

which has the eigenvectors as columns. By matrix multiplication

$$\begin{aligned}
TC &= T[\ \mathbf{r}_1 \quad \mathbf{r}_2 \quad \cdots \quad \mathbf{r}_m\ ] = [\ T\mathbf{r}_1 \quad T\mathbf{r}_2 \quad \cdots \quad T\mathbf{r}_m\ ] \\
&= [\ \lambda_1\mathbf{r}_1 \quad \lambda_2\mathbf{r}_2 \quad \cdots \quad \lambda_m\mathbf{r}_m\ ] \qquad \text{(by (4.15))} \\
&= [\ \mathbf{r}_1 \quad \mathbf{r}_2 \quad \cdots \quad \mathbf{r}_m\ ]
\begin{bmatrix}
\lambda_1 & 0 & \cdots & 0 \\
0 & \lambda_2 & \cdots & 0 \\
\vdots & \vdots & \ddots & \vdots \\
0 & 0 & \cdots & \lambda_m
\end{bmatrix} \\
&= CD, \tag{4.17}
\end{aligned}$$

where D is the diagonal matrix of eigenvalues defined by

$$D =
\begin{bmatrix}
\lambda_1 & 0 & \cdots & 0 \\
0 & \lambda_2 & \cdots & 0 \\
\vdots & \vdots & \ddots & \vdots \\
0 & 0 & \cdots & \lambda_m
\end{bmatrix}. \tag{4.18}$$

Hence if (4.17) is multiplied on the right by C^{-1}, then

$$T = CDC^{-1}.$$

Powers of T can now be easily found since

$$\begin{aligned}
T^2 &= (CDC^{-1})(CDC^{-1}) = (CD)(CC^{-1})(DC^{-1}) \\
&= (CD)I_m(DC^{-1}) = CD^2C^{-1}.
\end{aligned}$$

Similarly, for the general power n,

$$T^n = CD^nC^{-1}, \tag{4.19}$$

where

$$D^n =
\begin{bmatrix}
\lambda_1^n & 0 & \cdots & 0 \\
0 & \lambda_2^n & \cdots & 0 \\
\vdots & \vdots & \ddots & \vdots \\
0 & 0 & \cdots & \lambda_m^n
\end{bmatrix}.$$

---

[3] The notation A^t denotes the transpose of any matrix A in which rows and columns are interchanged.

Although Eqn (4.19) is a useful formula for T^n, the algebra involved can be very heavy even for quite modest chains with $m = 4$ or 5, particularly if T contains unknown constants or parameters. On the other hand, mathematical software is now very easy to apply for the numerical calculation of eigenvalues, eigenvectors, and powers of matrices for quite large systems.

As we shall see, the behaviour of T^n for large n will depend on the eigenvalues $\lambda_1, \lambda_2, \ldots, \lambda_n$: a limit may or may not exist.

Example 4.4 *Find the eigenvalues and eigenvectors of the stochastic matrix*

$$T = \begin{bmatrix} \frac{1}{4} & \frac{1}{2} & \frac{1}{4} \\ \frac{1}{2} & \frac{1}{4} & \frac{1}{4} \\ \frac{1}{4} & \frac{1}{4} & \frac{1}{2} \end{bmatrix}.$$

Construct a formula for T^n, and find $\lim_{n \to \infty} T^n$.

The eigenvalues of T are given by

$$
\begin{aligned}
\det(T - \lambda I_3) &= \begin{vmatrix} \frac{1}{4} - \lambda & \frac{1}{2} & \frac{1}{4} \\ \frac{1}{2} & \frac{1}{4} - \lambda & \frac{1}{4} \\ \frac{1}{4} & \frac{1}{4} & \frac{1}{2} - \lambda \end{vmatrix} \\[2mm]
&= \begin{vmatrix} 1 - \lambda & 1 - \lambda & 1 - \lambda \\ \frac{1}{2} & \frac{1}{4} - \lambda & \frac{1}{4} \\ \frac{1}{4} & \frac{1}{4} & \frac{1}{2} - \lambda \end{vmatrix} \qquad \text{(adding all the rows)} \\[2mm]
&= (1 - \lambda) \begin{vmatrix} 1 & 1 & 1 \\ \frac{1}{2} & \frac{1}{4} - \lambda & \frac{1}{4} \\ \frac{1}{4} & \frac{1}{4} & \frac{1}{2} - \lambda \end{vmatrix} \\[2mm]
&= (1 - \lambda) \begin{vmatrix} 1 & 1 & 1 \\ \frac{1}{2} & \frac{1}{4} - \lambda & \frac{1}{4} \\ 0 & 0 & \frac{1}{4} - \lambda \end{vmatrix} \quad (\text{ subtracting } \tfrac{1}{4} \text{ of row 1 from row 3}) \\[2mm]
&= (1 - \lambda)(\tfrac{1}{4} - \lambda)(-\tfrac{1}{4} - \lambda) = 0.
\end{aligned}
$$

Define the eigenvalues to be

$$\lambda_1 = 1, \qquad \lambda_2 = \tfrac{1}{4}, \qquad \lambda_3 = -\tfrac{1}{4}.$$

The eigenvectors $\mathbf{r}_i (i = 1, 2, 3)$ satisfy

$$(T - \lambda_i I_3)\mathbf{r}_i = \mathbf{0}, \qquad (i = 1, 2, 3).$$

Some routine calculations give

$$\mathbf{r}_1 = \begin{bmatrix} 1 \\ 1 \\ 1 \end{bmatrix}, \qquad \mathbf{r}_2 = \begin{bmatrix} -1 \\ -1 \\ 2 \end{bmatrix}, \qquad \mathbf{r}_3 = \begin{bmatrix} 1 \\ -1 \\ 0 \end{bmatrix}.$$

We now let

$$C = \begin{bmatrix} \mathbf{r}_1 & \mathbf{r}_2 & \mathbf{r}_3 \end{bmatrix} = \begin{bmatrix} 1 & -1 & 1 \\ 1 & -1 & -1 \\ 1 & 2 & 0 \end{bmatrix}.$$

Finally, from (4.19)

$$T^n = CD^nC^{-1}$$

$$= \frac{1}{6}\begin{bmatrix} 1 & -1 & 1 \\ 1 & -1 & -1 \\ 1 & 2 & 0 \end{bmatrix}\begin{bmatrix} 1 & 0 & 0 \\ 0 & (\frac{1}{4})^n & 0 \\ 0 & 0 & (-\frac{1}{4})^n \end{bmatrix}\begin{bmatrix} 2 & 2 & 2 \\ -1 & -1 & 2 \\ 3 & -3 & 0 \end{bmatrix}.$$

In this example two eigenvalues have magnitudes less than one so that T^n will approach a limit in which each row is the stationary distribution.

As $n \to \infty$,

$$T^n \to \frac{1}{6}\begin{bmatrix} 1 & -1 & 1 \\ 1 & -1 & -1 \\ 1 & 2 & 0 \end{bmatrix}\begin{bmatrix} 1 & 0 & 0 \\ 0 & 0 & 0 \\ 0 & 0 & 0 \end{bmatrix}\begin{bmatrix} 2 & 2 & 2 \\ -1 & -1 & 2 \\ 3 & -3 & 0 \end{bmatrix}$$

$$= \begin{bmatrix} \frac{1}{3} & \frac{1}{3} & \frac{1}{3} \\ \frac{1}{3} & \frac{1}{3} & \frac{1}{3} \\ \frac{1}{3} & \frac{1}{3} & \frac{1}{3} \end{bmatrix} = Q, \text{ say.}$$

Suppose now that T is the transition matrix of a 3-state Markov chain, and that the initial probability distribution is $\mathbf{p}^{(0)}$. Then the probability distribution after n steps is

$$\mathbf{p}^{(n)} = \mathbf{p}^{(0)}T^n.$$

The invariant probability distribution $\mathbf{p}$ is

$$\mathbf{p} = \lim_{n \to \infty} \mathbf{p}^{(n)} = \lim_{n \to \infty} \mathbf{p}^{(0)}T^n = \mathbf{p}^{(0)}Q = \begin{bmatrix} \frac{1}{3} & \frac{1}{3} & \frac{1}{3} \end{bmatrix}.$$

The vector $\mathbf{p}$ gives the long-term probability distribution across the three states. In other words, if any snapshot of the system is eventually taken for large n, then the system is equally likely (in this example) to lie in each of the states, and this is independent of the initial distribution $\mathbf{p}^{(0)}$.

This particular example is covered by the **Perron–Frobenius theorem**[4], which states: if T or T^r, for some r, is a **positive stochastic matrix** (that is, every element of T or T^r is strictly positive), then aside from the unit eigenvalue, all other eigenvalues have magnitudes less than one. (A proof is given by Grimmett and Stirzaker (1982).)

Absorbing barriers were referred to in Section 3.1 in the context of random walks. **Absorbing states** are recognizable in Markov chains by a value 1 in a diagonal element of the transition matrix. Since such matrices are stochastic, then all other elements in the same row must be zero. This means that once entered, there is no escape from absorbing state. For example, in the Markov chain with

$$T = \begin{bmatrix} \frac{1}{2} & \frac{1}{4} & \frac{1}{4} \\ 0 & 1 & 0 \\ \frac{1}{4} & \frac{1}{4} & \frac{1}{2} \end{bmatrix}, \tag{4.20}$$

then the state E_2 is absorbing.

As we illustrated in Section 3.1, diagrams showing transitions between states are

---

[4] Oskar Perron (1880–1975); Ferdinand Frobenius (1849–1917), German mathematicians.

particularly helpful. The states are represented by dots with linking directed curves or edges if a transition is possible: if no transition is possible then no directed edge is drawn. Thus the transition diagram for the three-state chain with transition matrix given by Eqn (4.20) is shown in Figure 4.1. In graph theory terminology, Figure 4.1 shows a **directed graph**. It can be seen that once entered, there is no escape from the absorbing state E_2.

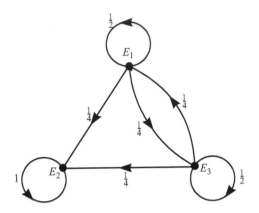

Figure 4.1 *Transition diagram for the transition matrix in Eqn (4.20).*

The eigenvalues of T given by Eqn (4.20) are $\lambda_1 = 1$, $\lambda_2 = \frac{1}{4}$, $\lambda_3 = \frac{3}{4}$. (The eigenvalues of T satisfy the conclusions of the Perron–Frobenius theorem above, but T in this case is not a positive matrix, which indicates a more general version of the theorem: see the reference above again.) The corresponding matrix of eigenvectors is

$$C = \begin{bmatrix} 1 & -1 & 0 \\ 1 & 0 & 0 \\ 1 & 1 & 1 \end{bmatrix}.$$

Using the method illustrated by Example 4.3, it follows that

$$
\begin{aligned}
T^n &= CD^nC^{-1} \\
&= \begin{bmatrix} 1 & -1 & 1 \\ 1 & 0 & 0 \\ 1 & 1 & 1 \end{bmatrix}
\begin{bmatrix} 1 & 0 & 0 \\ 0 & (\frac{1}{4})^n & 0 \\ 0 & 0 & (\frac{3}{4})^n \end{bmatrix}
\begin{bmatrix} 0 & 1 & 0 \\ -\frac{1}{2} & 0 & \frac{1}{2} \\ \frac{1}{2} & -1 & \frac{1}{2} \end{bmatrix} \\
&\to \begin{bmatrix} 1 & -1 & 1 \\ 1 & 0 & 0 \\ 1 & 1 & 1 \end{bmatrix}
\begin{bmatrix} 1 & 0 & 0 \\ 0 & 0 & 0 \\ 0 & 0 & 0 \end{bmatrix}
\begin{bmatrix} 0 & 1 & 0 \\ -\frac{1}{2} & 0 & \frac{1}{2} \\ \frac{1}{2} & -1 & \frac{1}{2} \end{bmatrix} \\
&= \begin{bmatrix} 0 & 1 & 0 \\ 0 & 1 & 0 \\ 0 & 1 & 0 \end{bmatrix} = Q,
\end{aligned}
$$

say, as $n \to \infty$. This implies that

$$\mathbf{p} = \lim_{n \to \infty} \mathbf{p}^{(0)} T^n = \mathbf{p}^{(0)} Q = \begin{bmatrix} 0 & 1 & 0 \end{bmatrix}$$

for any initial distribution $\mathbf{p}^{(0)}$. This means that the system ultimately ends in E_2 with probability 1.

Example 4.5 *(An illness–death model) A possible simple illness–death model can be repre-sented by a four-state Markov chain in which E_1 is a state in which an individual is free of a particular disease, E_2 is a state in which the individual has the disease, and E_3 and E_4 are respectively death states arising as a consequence of death as a result of the disease, or from other causes. During some appropriate time interval (perhaps an annual cycle), we assign probabilities to the transition between the states. Suppose that the transition matrix is (in the order of the states),*

$$ T = \begin{bmatrix} \frac{1}{2} & \frac{1}{4} & 0 & \frac{1}{4} \\ \frac{1}{4} & \frac{1}{2} & \frac{1}{8} & \frac{1}{8} \\ 0 & 0 & 1 & 0 \\ 0 & 0 & 0 & 1 \end{bmatrix}. \tag{4.21} $$

Find the probabilities that a person ultimately dies after a large number of transitions from the disease given that he or she did not have the disease initially.

As we might expect, this Markov chain has two absorbing states, E_3 and E_4. The individual probabilities can be interpreted as follows: for example, $p_{11} = \frac{1}{2}$ means that an individual, given that s/he is free of the disease in a certain period, has probability $\frac{1}{2}$ of remaining free of the disease; $p_{24} = \frac{1}{8}$ means that the probability that an individual has the disease but dies from other causes is $\frac{1}{8}$, and so on.

The transition diagram is shown in Figure 4.2.

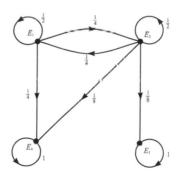

Figure 4.2 *Transition diagram for the illness–death model.*

In this example it is simpler to **partition** the matrix T as follows. Let

$$ T = \begin{bmatrix} A & B \\ O_{22} & I_2 \end{bmatrix}, $$

where the **submatrices**[5] A and B are given by

$$ A = \begin{bmatrix} \frac{1}{2} & \frac{1}{4} \\ \frac{1}{4} & \frac{1}{2} \end{bmatrix}, \quad B = \begin{bmatrix} 0 & \frac{1}{4} \\ \frac{1}{8} & \frac{1}{8} \end{bmatrix}, \quad O_{22} = \begin{bmatrix} 0 & 0 \\ 0 & 0 \end{bmatrix}, \quad I_2 = \begin{bmatrix} 1 & 0 \\ 0 & 1 \end{bmatrix}. $$

---

[5] A submatrix of a matrix is a matrix obtained by deleting any collection of rows and columns in the original matrix.

Note that A and B are not stochastic matrices. We then observe that

$$T^2 = \begin{bmatrix} A & B \\ O_{22} & I_2 \end{bmatrix} \begin{bmatrix} A & B \\ O_{22} & I_2 \end{bmatrix} = \begin{bmatrix} A^2 & (A+I_2)B \\ O_{22} & I_2 \end{bmatrix},$$

$$T^3 = \begin{bmatrix} A^3 & (A^2+A+I_2)B \\ O_{22} & I_2 \end{bmatrix},$$

and, in general,

$$T^n = \begin{bmatrix} A^n & (I_2+A+A^2+\cdots+A^{n-1})B \\ O_{22} & I_2 \end{bmatrix}.$$

Now let

$$S_n = I_2 + A + \cdots + A^{n-1}.$$

It then follows that

$$(I_2 - A)S_n = (I_2 + A + \cdots + A^{n-1}) - (A + A^2 + \cdots + A^n) = I_2 - A^n,$$

so that

$$S_n = (I_2 - A)^{-1}(I_2 - A^n).$$

Hence,

$$T^n = \begin{bmatrix} A^n & (I_2-A)^{-1}(I_2-A^n)B \\ O_{22} & I_2 \end{bmatrix}.$$

In the matrix T, A is *not* a stochastic matrix but the method of Section 4.3 to find A^n still works. The eigenvalues of A are given by

$$\det(A - \lambda I_2) = \begin{vmatrix} \frac{1}{2}-\lambda & \frac{1}{4} \\ \frac{1}{4} & \frac{1}{2}-\lambda \end{vmatrix} = (\lambda - \tfrac{1}{4})(\lambda - \tfrac{3}{4}) = 0.$$

Let $\lambda_1 = \frac{1}{4}$ and $\lambda_2 = \frac{3}{4}$ (note that there is not a unit eigenvalue in this case since A is not row-stochastic). The matrix C of eigenvectors is

$$C = \begin{bmatrix} -1 & 1 \\ 1 & 1 \end{bmatrix}.$$

Hence,

$$A^n = \frac{1}{2} \begin{bmatrix} -1 & 1 \\ 1 & 1 \end{bmatrix} \begin{bmatrix} (\frac{1}{4})^n & 0 \\ 0 & (\frac{3}{4})^n \end{bmatrix} \begin{bmatrix} -1 & 1 \\ 1 & 1 \end{bmatrix}.$$

For the submatrix A, $A^n \to 0$ as $n \to \infty$, so that

$$T^n \to \begin{bmatrix} O_{22} & (I_2-A)^{-1}B \\ O_{22} & I_2 \end{bmatrix} = \begin{bmatrix} 0 & 0 & \frac{1}{6} & \frac{5}{6} \\ 0 & 0 & \frac{1}{3} & \frac{2}{3} \\ 0 & 0 & 1 & 0 \\ 0 & 0 & 0 & 1 \end{bmatrix} = Q,$$

say. In this case

$$\mathbf{p} = \begin{bmatrix} 0 & 0 & \frac{1}{6}p_1^{(0)} + \frac{1}{3}p_2^{(0)} + p_3^{(0)} & \frac{5}{6}p_1^{(0)} + \frac{2}{3}p_2^{(0)} + p_4^{(0)} \end{bmatrix}.$$

This is a **limiting distribution** which depends on the initial distribution. For example, if $\mathbf{p}^{(0)} = \begin{bmatrix} 1 & 0 & 0 & 0 \end{bmatrix}$, then

$$\mathbf{p} = \begin{bmatrix} 0 & 0 & \frac{1}{6} & \frac{5}{6} \end{bmatrix},$$

from which it can be seen that the probability of an individual not having the disease initially but ultimately dying as a result of the disease is $\frac{1}{6}$. On the other hand $\mathbf{p}^{(0)} = \begin{bmatrix} \frac{1}{2} & 0 & \frac{1}{2} & 0 \end{bmatrix}$ will give the different limit

$$\mathbf{p} = \begin{bmatrix} 0 & 0 & \frac{7}{12} & \frac{5}{12} \end{bmatrix}.$$

The examples we have considered so far have stochastic matrices which have *real* eigenvalues. How do we proceed if the eigenvalues are complex? Consider the following example.

Example 4.6 *Find the eigenvalues and eigenvectors of the stochastic matrix*

$$T = \begin{bmatrix} \frac{1}{2} & \frac{1}{8} & \frac{3}{8} \\ 1 & 0 & 0 \\ \frac{1}{4} & \frac{1}{2} & \frac{1}{4} \end{bmatrix}.$$

Construct a formula for T^n, and find $\lim_{n\to\infty} T^n$.

The eigenvalues of T are given by

$$
\begin{aligned}
\det(T - \lambda I_3) &= \begin{vmatrix} \frac{1}{2} - \lambda & \frac{1}{8} & \frac{3}{8} \\ 1 & -\lambda & 0 \\ \frac{1}{4} & \frac{1}{2} & \frac{1}{4} - \lambda \end{vmatrix} \\
&= (1 - \lambda) \begin{vmatrix} 1 & \frac{1}{8} & \frac{3}{8} \\ 1 & -\lambda & 0 \\ 1 & \frac{1}{2} & \frac{1}{4} - \lambda \end{vmatrix} \quad \text{(adding all the columns)} \\
&= (1 - \lambda) \begin{vmatrix} 0 & \frac{1}{8} + \lambda & \frac{3}{8} \\ 1 & -\lambda & 0 \\ 0 & \frac{1}{2} + \lambda & \frac{1}{4} - \lambda \end{vmatrix} \quad \text{(subtracting row 2 from rows 1 and 3)} \\
&= -(1 - \lambda) \begin{vmatrix} \frac{1}{8} + \lambda & \frac{3}{8} \\ \frac{1}{2} + \lambda & \frac{1}{4} - \lambda \end{vmatrix} \\
&= (1 - \lambda)(\lambda^2 + \tfrac{1}{4}\lambda + \tfrac{5}{32}).
\end{aligned}
$$

Hence the eigenvalues are

$$\lambda_1 = 1, \qquad \lambda_2 = \tfrac{1}{8}(-1 + 3i), \qquad \lambda_3 = \tfrac{1}{8}(-1 - 3i),$$

two of which are complex conjugates. However, we proceed to find the (complex) eigenvectors which are given, as before, by

$$(T - \lambda_i I_3)\mathbf{r}_i = \mathbf{0}, \qquad (i = 1, 2, 3).$$

They can be chosen as

$$\mathbf{r}_1 = \begin{bmatrix} 1 \\ 1 \\ 1 \end{bmatrix}, \qquad \mathbf{r}_2 = \begin{bmatrix} -7 - 9i \\ -16 + 24i \\ 26 \end{bmatrix}, \qquad \mathbf{r}_3 = \begin{bmatrix} -7 + 9i \\ -16 - 24i \\ 26 \end{bmatrix}.$$

We now define C by

$$C = \begin{bmatrix} \mathbf{r}_1 & \mathbf{r}_2 & \mathbf{r}_3 \end{bmatrix} = \begin{bmatrix} 1 & -7 - 9i & -7 + 9i \\ 1 & -16 + 24i & -16 - 24i \\ 1 & 26 & 26 \end{bmatrix}.$$

Inevitably the algebra becomes heavier for the case of complex roots. The inverse of C is given by

$$C^{-1} = \frac{1}{780} \begin{bmatrix} 416 & 156 & 208 \\ -8 + 14i & -3 - 11i & 11 - 3i \\ -8 - 14i & -3 - 11i & 11 + 3i \end{bmatrix}.$$

The diagonal matrix D of eigenvalues becomes

$$D = \begin{bmatrix} 1 & 0 & 0 \\ 0 & (\frac{-1+3i}{8})^n & 0 \\ 0 & 0 & (\frac{-1-3i}{8})^n \end{bmatrix} \rightarrow \begin{bmatrix} 1 & 0 & 0 \\ 0 & 0 & 0 \\ 0 & 0 & 0 \end{bmatrix}$$

as $n \to \infty$ since $|(-1 \pm 3i)/8| = \sqrt{10}/8 < 1$. Finally, T^n can be calculated from the formula

$$T^n = CD^nC^{-1}.$$

As $n \to \infty$,

$$T^n \rightarrow \begin{bmatrix} \frac{8}{15} & \frac{1}{5} & \frac{4}{15} \\ \frac{8}{15} & \frac{1}{5} & \frac{4}{15} \\ \frac{8}{15} & \frac{1}{5} & \frac{4}{15} \end{bmatrix} = Q.$$

We conclude that complex eigenvalues can be dealt with by the same procedure as real ones: the resulting formulas for T^n and its limit Q will turn out to be real matrices.

This example illustrates the Perron–Frobenius theorem quoted previously in this section. Although T is not positive, T^2 is a positive matrix. It can be confirmed that, for the complex eigenvalues

$$\lambda_1 = 1 \qquad |\lambda_2| = |\lambda_3| = \frac{\sqrt{10}}{8} < 1.$$

We have not considered the cases of repeated eigenvalues. The matrix algebra required is beyond the scope of this textbook. However, Problems 4.8 and 4.9 at the end of this chapter suggest three such cases which are worth investigating.

4.5 Gambler's ruin as a Markov chain

We start by summarizing the game (see Section 2.1). It is a game of chance between two players A and B. The gambler A starts with k units (pounds or dollars, etc.) and the opponent B with $a - k$ units, where a and k are integers. At each play, A either wins from B one unit with probability p or loses one unit to B with probability $q = 1 - p$. The game ends when either player A or B has no stake. What is the probability that A loses?

The states are $E_0, E_1, \ldots, E_a$ (it is convenient in this application to let the list run from 0 to a), where E_r is the state in which the gambler A has r units. This is a Markov chain, but the transitions are only possible between neighbouring states. This is also the case for the *simple* random walk. We interpret E_0 and E_a as absorbing states since the game ends when these states are reached. From the rules of the game, the transition matrix is

$$
T = \begin{bmatrix}
1 & 0 & 0 & \cdots & 0 & 0 & 0 \\
1-p & 0 & p & \cdots & 0 & 0 & 0 \\
\vdots & \vdots & \vdots & \ddots & \vdots & \vdots & \vdots \\
0 & 0 & 0 & \cdots & 1-p & 0 & p \\
0 & 0 & 0 & \cdots & 0 & 0 & 1
\end{bmatrix} \quad \text{(an } (a+1) \times (a+1) \text{ matrix).} \quad (4.22)
$$

T is an example of a **tridiagonal matrix**.

The diagonalization of T is not really a practical proposition for this $(a+1) \times (a+1)$ matrix except for small a. This is often the case for chains with a large number of states.

The initial distribution vector $\mathbf{p}^{(0)}$ for the gambler's ruin problem has the elements

$$
p_i^{(0)} = \begin{cases} 0 & i \neq k, \\ 1 & i = k, \end{cases} \quad (0 \leq i \leq a, \quad 1 \leq k \leq a-1),
$$

assuming an initial stake of k units. In the vector $\mathbf{p}^{(n)}$, the component $p_0^{(n)}$ is the probability that the gambler loses the game by the nth play, and $p_a^{(n)}$ is the probability that the gambler has won by the nth play.

Example 4.7 *In a gambler's ruin problem suppose that p is the probability that the gambler wins at each play, and that $a = 5$ and that the gambler's initial stake is $k = 3$ units. Compute the probability that the gambler loses/wins by the fourth play. What is the probability that the gambler actually wins at the fourth play?*

In this example, the transition matrix Eqn (4.21) is the 6×6 matrix

$$
T = \begin{bmatrix}
1 & 0 & 0 & 0 & 0 & 0 \\
1-p & 0 & p & 0 & 0 & 0 \\
0 & 1-p & 0 & p & 0 & 0 \\
0 & 0 & 1-p & 0 & p & 0 \\
0 & 0 & 0 & 1-p & 0 & p \\
0 & 0 & 0 & 0 & 0 & 1
\end{bmatrix},
$$

and

$$
\mathbf{p}^{(0)} = \begin{bmatrix} 0 & 0 & 0 & 1 & 0 & 0 \end{bmatrix}.
$$

Then

$$
\begin{aligned}
\mathbf{p}^{(4)} &= \mathbf{p}^{(0)} T^4 \\
&= \begin{bmatrix} (1-p)^3 & 3(1-p)^3 & 0 & 5(1-p)^2 p^2 & 0 & p^2 + 2(1-p)p^3 \end{bmatrix}.
\end{aligned}
$$

Symbolic computation is helpful for the matrix powers. Alternatively the distribution can be obtained using probability arguments. The probability that the gambler loses by the fourth play is

$$
p_0^{(4)} = (1-p)^3,
$$

which can only occur as the probability of three successive losses. The probability that the gambler wins at the fourth play is

$$
p_5^{(4)} = p^2 + 2(1-p)p^3,
$$

derived from two successive wins, or from the sequences win/lose/win/win or lose/win/win/win which together give the term $2(1-p)p^3$. The probability that the gambler actually wins *at* the fourth play is $2(1-p)p^3$.

For the general $(a+1) \times (a+1)$ transition matrix (4.22) for the gambler's ruin, whilst T^n is difficult to obtain, $Q = \lim_{n \to} T^n$ can be found if we make a reasonable assumption that the limiting matrix is of the form

$$
Q = \begin{bmatrix}
1 & 0 & \cdots & 0 & 0 \\
u_1 & 0 & \cdots & 0 & 1-u_1 \\
u_2 & 0 & \cdots & 0 & 1-u_2 \\
\vdots & \vdots & \ddots & \vdots & \vdots \\
u_{a-1} & 0 & \cdots & 0 & 1-u_{a-1} \\
0 & 0 & \cdots & 0 & 1
\end{bmatrix},
$$

where $u_1, u_2, \ldots, u_{a-1}$ are to be determined. The implication of the zeros in columns 2 to $a-1$ is that eventually the chain must end in either state E_0 or E_a (see Section 2.2). We then observe that, after rearranging indices,

$$Q = \lim_{n \to \infty} T^{n+1} = (\lim_{n \to \infty} T^n)T = QT, \tag{4.23}$$

and

$$Q = \lim_{n \to \infty} T^{n+1} = T(\lim_{n \to \infty} T^n) = TQ. \tag{4.24}$$

In this case (4.23) turns out to be an identity. However, (4.24) implies

$0 = TQ - Q = (T - I_{a+1})Q$

$$
= \begin{bmatrix}
0 & 0 & 0 & \cdots & 0 & 0 & 0 \\
1-p & -1 & p & \cdots & 0 & 0 & 0 \\
0 & 1-p & -1 & \cdots & 0 & 0 & 0 \\
\vdots & \vdots & \vdots & \ddots & \vdots & \vdots & \vdots \\
0 & 0 & 0 & \cdots & 1-p & -1 & p \\
0 & 0 & 0 & \cdots & 0 & 0 & 0
\end{bmatrix}
\begin{bmatrix}
1 & 0 & \cdots & 0 & 0 \\
u_1 & 0 & \cdots & 0 & 1-u_1 \\
u_2 & 0 & \cdots & 0 & 1-u_2 \\
\vdots & \vdots & \ddots & \vdots & \vdots \\
u_{a-1} & 0 & \cdots & 0 & 1-u_{a-1} \\
0 & 0 & \cdots & 0 & 1
\end{bmatrix}
$$

$$
= \begin{bmatrix}
0 & 0 & \cdots & 0 & 0 \\
(1-p) - u_1 + pu_2 & 0 & \cdots & 0 & -(1-p) + u_1 - pu_2 \\
(1-p)u_1 - u_2 + pu_3 & 0 & \cdots & 0 & -(1-p)u_1 + u_2 - pu_3 \\
\vdots & \vdots & \ddots & \vdots & \vdots \\
(1-p)u_{a-2} - u_{a-1} & 0 & \cdots & 0 & -(1-p)u_{a-2} + u_{a-1} \\
0 & 0 & \cdots & 0 & 0
\end{bmatrix}.
$$

The right-hand side is zero if the elements in the first and last columns are all zero. However, note that the elements in the last column duplicate the ones in the first column. The result is the set of difference equations

$$pu_2 - u_1 + (1-p) = 0,$$

$$pu_{k+2} - u_{k+1} + (1-p)u_k = 0, \qquad (k = 1, 2, \ldots, a-3),$$

$$-u_{a-1} + (1-p)u_{a-2} = 0.$$

This is equivalent to writing

$$pu_{k+2} - u_{k+1} + (1-p)u_k = 0, \qquad (k = 0, 1, 2, \ldots, a-2),$$

$$u_0 = 1, \qquad u_a = 0,$$

the latter two equations now defining the boundary conditions. These are the difference equations and boundary conditions for the gambler's ruin problem derived using the law of total probability in Eqns (2.2) and (2.3).

4.6 Classification of states

Let us return to the general m-state chain with states $E_1, E_2, \ldots, E_m$ and transition matrix

$$T = [p_{ij}], \qquad (1 \leq i, j \leq m).$$

For a homogeneous chain, recollect that p_{ij} is the probability that a transition occurs between E_i and E_j at any step or change of state in the chain. We intend to investigate and classify some of the more common types of states which can occur in Markov chains. This will be a brief treatment, using mainly examples of what is an extensive algebraic subject.

(a) Absorbing state

We have already met one type of state—namely the **absorbing state** (see Section 3.1). Once entered there is no escape from an absorbing state. An absorbing state E_i is characterized by the probabilities

$$p_{ii} = 1, \qquad p_{ij} = 0, (j \neq i, \quad j = 1, 2, \ldots, m), \tag{4.25}$$

in the i-th row of T.

(b) Periodic state

The probability of a return to E_i at step n is $p_{ii}^{(n)}$. Let t be an integer greater than 1. Suppose that

$$p_{ii}^{(n)} = 0 \text{ for } n \neq t, 2t, 3t, \ldots$$

$$p_{ii}^{(n)} \neq 0 \text{ for } n = t, 2t, 3t, \ldots.$$

In this case the state E_i is said to be **periodic** with period t. If, for a state, no such t exists with this property, then the state is described as **aperiodic**.

Let

$$d(i) = \gcd\{n | p_{ii}^{(n)} > 0\},$$

that is, the greatest common divisor (gcd)[6] of the set of integers n for which $p_{ii}^{(n)} > 0$. Then the state E_i is said to be **periodic** if $d(i) > 1$ and **aperiodic** if $d(i) = 1$. This definition, which includes the earlier case, covers the appearance of *delayed* periodicity in which a sequence of initial terms are zero (see Problem 4.12 for an

---

[6] $\gcd\{n | p_{ii}^{(n)} > 0\}$ is the largest integer which divides all $p_{ii}^{(n)} > 0$ exactly.

example of such periodic states which satisfy this general definition but not the one above) .

Example 4.8. *A four-state Markov chain has the transition matrix*

$$T = \begin{bmatrix} 0 & \frac{1}{2} & 0 & \frac{1}{2} \\ 0 & 0 & 1 & 0 \\ 1 & 0 & 0 & 0 \\ 0 & 0 & 1 & 0 \end{bmatrix}.$$

Show that all states have period 3.

The transition diagram is shown in Figure 4.3, from which it is clear that all states have period 3. For example, if the chain starts in E_1, then returns to E_1 are only possible at steps

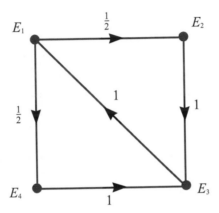

Figure 4.3 *The transition diagram for Example 4.8.*

$3, 6, 9, \ldots$ either through E_2 or E_3.

The analysis of chains with periodic states can be complicated. However, one can check for a suspected periodicity as follows. By direct computation

$$S = T^3 = \begin{bmatrix} 1 & 0 & 0 & 0 \\ 0 & \frac{1}{2} & 0 & \frac{1}{2} \\ 0 & 0 & 1 & 0 \\ 0 & \frac{1}{2} & 0 & \frac{1}{2} \end{bmatrix}.$$

In this example,

$$S^2 = T^6 = SS = S,$$

so that

$$S^r = T^{3r} = S, \qquad (r = 1, 2, \ldots),$$

which always has nonzero elements on its diagonal. On the other hand,

$$S^{r+1} = S^r S = \begin{bmatrix} 0 & \frac{1}{2} & 0 & \frac{1}{2} \\ 0 & 0 & 1 & 0 \\ 1 & 0 & 0 & 0 \\ 0 & 0 & 1 & 0 \end{bmatrix}, \qquad S^{r+2} = S^r S^2 = \begin{bmatrix} 0 & 0 & 1 & 0 \\ 1 & 0 & 0 & 0 \\ 0 & \frac{1}{2} & 0 & \frac{1}{2} \\ 1 & 0 & 0 & 0 \end{bmatrix},$$

and both these matrices have zero diagonal elements for $r = 1, 2, 3, \ldots$. Hence, for $i = 1, 2, 3, 4$,

$$p_{ii}^{(n)} = 0 \text{ for } n \neq 3, 6, 9, \ldots,$$

$$p_{ii}^{(n)} \neq 0 \text{ for } n = 3, 6, 9, \ldots,$$

which means that all states are period 3. In this example

$$d(i) = \gcd\{3, 6, 9, \ldots\} = 3, \text{ for } i = 1, 2, 3, 4.$$

(c) Persistent state

Let $f_j^{(n)}$ be the probability that the **first return** or **visit**[7] to E_j occurs at the n-th step. This probability is not the same as $p_{jj}^{(n)}$, which is the probability that a return occurs at the n-th step, and includes possible returns at steps $1, 2, 3, \ldots, n-1$ also. It follows that

$$p_{jj}^{(1)} (= p_{jj}) - f_j^{(1)}, \tag{4.26}$$

$$p_{jj}^{(2)} = f_j^{(2)} + f_j^{(1)} p_{jj}^{(1)}, \tag{4.27}$$

$$p_{jj}^{(3)} = f_j^{(3)} + f_j^{(1)} p_{jj}^{(2)} + f_j^{(2)} p_{jj}^{(1)}, \tag{4.28}$$

and, in general,

$$p_{jj}^{(n)} = f_j^{(n)} + \sum_{r-1}^{n-1} f_j^{(r)} p_{jj}^{(n-r)} \qquad (n \geq 2). \tag{4.29}$$

Think about the meaning of these equations: for example, (4.27) states that the probability of a return to E_j at step 2 is the sum of the probability of a first return at step 2 plus the probability of a first return at step 1. The terms in Eqn (4.28) imply that the probability of a return at the third step is the probability of a first return at the third step, or the probability of a first return at the first step and a return two steps later, or the probability of a first return at the second step and a return one step later.

Equations (4.26) and (4.29) become iterative formulas for the sequence of first returns $f_j^{(n)}$, which can be expressed as:

$$f_j^{(1)} = p_{jj}, \tag{4.30}$$

$$f_j^{(n)} = p_{jj}^{(n)} - \sum_{r=1}^{n-1} f_j^{(r)} p_{jj}^{(n-r)} \qquad (n \geq 2). \tag{4.31}$$

The probability that a chain returns at some step to the state E_j is

$$f_j = \sum_{n=1}^{\infty} f_j^{(n)}.$$

If $f_j = 1$, then a return to E_j is certain, and E_j is called a **persistent state**.

---

[7] We first met first returns in Section 3.4.

Example 4.9 *A three-state Markov chain has the transition matrix*

$$T = \begin{bmatrix} p & 1-p & 0 \\ 0 & 0 & 1 \\ 1-q & 0 & q \end{bmatrix},$$

where $0 < p < 1, 0 < q < 1$. *Show that the state* E_1 *is persistent.*

For simple chains a direct approach using the transition diagram is often easier than the formula (4.29) for $f_j^{(n)}$. For this example the transition diagram is shown in Figure 4.4. If a

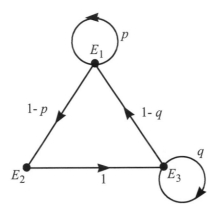

Figure 4.4 *The transition diagram for Example 4.9.*

sequence starts in E_1, then it can be seen that *first returns* to E_1 can be made to E_1 at every step except for $n = 2$, since after two steps the chain must be in state E_3. From the figure it can be argued that

$$f_1^{(1)} = p, \qquad f_1^{(2)} = 0, \qquad f_1^{(3)} = (1-p) \cdot 1 \cdot (1-q),$$

$$f_1^{(n)} = (1-p).1.q^{n-3}.(1-q), \quad (n \geq 4).$$

The last result for $f_1^{(n)}$ for $n \geq 4$ follows from the following sequence of transitions:

$$E_1 \ E_2 \ \overbrace{E_3 \ E_3 \ \cdots \ E_3}^{(n-3) \text{ times}} \ E_1.$$

The probability f_1 that the system returns at least once to E_1 is

$$
\begin{aligned}
f_1 = \sum_{n=1}^{\infty} f_1^{(n)} &= p + \sum_{n=3}^{\infty}(1-p)(1-q)q^{n-3}, \\
&= p + (1-p)(1-q)\sum_{s=0}^{\infty} q^s, \qquad (s = n-3) \\
&= p + (1-p)\frac{(1-q)}{(1-q)} = 1,
\end{aligned}
$$

using the sum formula for the geometric series. Hence $f_1 = 1$, and consequently the state E_1 is persistent.

In fact all states are persistent. Note also that T^3 is a positive matrix.

The **mean recurrence time** μ_j of a persistent state E_j, for which $\sum_{n=1}^{\infty} f_j^{(n)} = 1$, is given by

$$\mu_j = \sum_{n=1}^{\infty} n f_j^{(n)}. \qquad (4.32)$$

In Example 4.7 above, the state E_1 is persistent and its mean recurrence time is given by

$$
\begin{aligned}
\mu_1 = \sum_{n=1}^{\infty} n f_1^{(n)} &= p + (1-p)(1-q) \sum_{n=3}^{\infty} n q^{n-3} \\
&= p + (1-p)(1-q) \left[\frac{3-2q}{(1-q)^2} \right] = \frac{3-2p-2q+pq}{1-q},
\end{aligned}
$$

which is finite. For some chains, however, the mean recurrence time can be infinite.

A persistent state E_j is said to be **null** if $\mu_j = \infty$, and **nonnull** if $\mu_j < \infty$. All states in Example 4.7 are nonnull persistent.

To create a simple example of a finite chain with a persistent null state, we consider a chain in which the transition probabilities depend on the step number n. Null states are not possible in finite chains with a *constant* transition matrix, but a proof will not be given here.

Example 4.10 *A three-state inhomogeneous Markov chain has the transition matrix*

$$
T_n = \begin{bmatrix} \frac{1}{2} & \frac{1}{2} & 0 \\ 0 & 0 & 1 \\ 1/(n+1) & 0 & n/(n+1) \end{bmatrix},
$$

where T_n is the transition matrix at step n. Show that E_1 is a persistent null state.

The transition diagram at a general step n is shown in Figure 4.5. From the figure

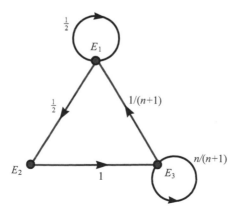

Figure 4.5 *The transition diagram for Example 4.10.*

$$f_1^{(1)} = \frac{1}{2}, \qquad f_1^{(2)} = 0, \qquad f_1^{(3)} = \frac{1}{2} \cdot 1 \cdot \frac{1}{4},$$

$$f_1^{(n)} = \frac{1}{2} \cdot 1 \cdot \frac{3}{4} \cdot \frac{4}{5} \cdots \frac{n-1}{n} \cdot \frac{1}{n+1} = \frac{3}{2n(n+1)}, \qquad (n \geq 4).$$

Hence,

$$f_1 = \frac{1}{2} + \frac{1}{8} + \frac{3}{2} \sum_{n=4}^{\infty} \frac{1}{n(n+1)}.$$

Since

$$\frac{1}{n(n+1)} = \frac{1}{n} - \frac{1}{n+1},$$

it follows that

$$\sum_{n=4}^{\infty} \frac{1}{n(n+1)} = \lim_{N \to \infty} \sum_{n=4}^{N} \left(\frac{1}{n} - \frac{1}{n+1} \right) = \lim_{N \to \infty} \left(\frac{1}{4} - \frac{1}{N+1} \right) = \frac{1}{4}.$$

Hence,

$$f_1 = \frac{5}{8} + \frac{3}{8} = 1,$$

which means that E_1 is persistent. On the other hand, the mean recurrence time

$$\mu_1 = \sum_{n=1}^{\infty} n f_1^{(n)} = \frac{7}{8} + \frac{3}{2} \sum_{n=4}^{\infty} \frac{n}{n(n+1)},$$

$$= \frac{7}{8} + \frac{3}{2} \left(\frac{1}{5} + \frac{1}{6} + \frac{1}{7} + \cdots \right) = \frac{7}{8} + \frac{3}{2} \sum_{n=5}^{\infty} \frac{1}{n}.$$

The series in the previous equation is the **harmonic series**

$$\sum_{n=1}^{\infty} \frac{1}{n} = 1 + \frac{1}{2} + \frac{1}{3} + \cdots,$$

minus the first four terms. The harmonic series is a well-known *divergent series*, which means that $\mu_1 = \infty$. Hence E_1 is persistent and null.

States can be both persistent and periodic. In the four-state chain with transition matrix

$$T = \begin{bmatrix} 0 & 1 & 0 & 0 \\ 0 & 0 & 0 & 1 \\ 0 & 1 & 0 & 0 \\ \frac{1}{2} & 0 & \frac{1}{2} & 0 \end{bmatrix},$$

all states are period 3, persistent, and non-null.

(d) Transient state

For a persistent state the probability of a first return at some step in the future is certain. For some states,

$$f_j = \sum_{n=1}^{\infty} f_j^{(n)} < 1, \tag{4.33}$$

which means that the probability of a first return is *not* certain. Such states are described as **transient**.

Example 4.11 *A four-state Markov chain has the transition matrix*

$$T = \begin{bmatrix} 0 & \frac{1}{2} & \frac{1}{4} & \frac{1}{4} \\ \frac{1}{2} & \frac{1}{2} & 0 & 0 \\ 0 & 0 & 1 & 0 \\ 0 & 0 & \frac{1}{2} & \frac{1}{2} \end{bmatrix}.$$

Show that E_1 is a transient state.

The transition diagram is shown in Figure 4.6. From the figure

$$f_1^{(1)} = 0, \quad f_1^{(2)} = \frac{1}{2} \cdot \frac{1}{2} = (\frac{1}{2})^2, \quad f_1^{(3)} = (\frac{1}{2})^3, \quad f_1^{(n)} = (\frac{1}{2})^n.$$

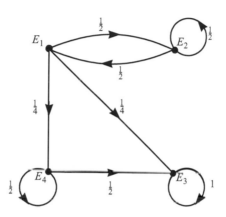

Figure 4.6 *The transition diagram for Example 4.11.*

Hence,

$$f_1 = \sum_{n=1}^{\infty} f_1^{(n)} - \sum_{n=2}^{\infty} (\tfrac{1}{2})^n = \tfrac{1}{2} < 1,$$

implying that E_1 is a transient state. The reason for the transience of E_1 can be seen from Figure 4.6, where transitions from E_3 or E_4 to E_1 or E_2 are not possible.

(e) Ergodic state

An important state which we will return to in the next section is the state which is persistent, nonnull, and aperiodic. This state is called **ergodic**[8]. Ergodic states are important in the classification of chains, and in the existence of limiting probability distributions, as we shall see in the following section.

Example 4.12 *In Example 4.9 we considered the three-state Markov chain with transition matrix*

$$T = \begin{bmatrix} p & 1-p & 0 \\ 0 & 0 & 1 \\ 1-q & 0 & q \end{bmatrix}$$

---

[8] Generally, the word *ergodic* means that a state will eventually return to a previous state, or put alternatively a sequence or sample in a state is representative of the whole chain.

where $0 < p < 1$, $0 < q < 1$. Show that state E_1 is ergodic.

It was shown in Example 4.9 that E_1 was persistent with

$$f_1^{(1)} = p, \qquad f_1^{(2)} = 0, \qquad f_1^{(n)} = (1-p)(1-q)q^{n-3}, \quad (n \geq 3).$$

It follows that its mean recurrence time is

$$\mu_1 = \sum_{n=1}^{\infty} n f_1^{(n)} = p + (1-p)(1-q) \sum_{n=3}^{\infty} n q^{n-3} = \frac{3 - 2q}{(1-q)^2} < \infty.$$

The convergence of μ_1 implies that E_1 is nonnull. Also the diagonal elements $p_{ii}^{(n)} > 0$ for $n \geq 3$ and $i = 1, 2, 3$, which means that E_1 is aperiodic. Hence from the definition above, E_1 (and E_2 and E_3 also) is ergodic.

4.7 Classification of chains

In the previous section we considered some defining properties of individual states. In this section we discuss properties of chains.

(a) Irreducible chains

An **irreducible chain** is one in which every state can be **reached** or is **accessible** from every other state in the chain in a finite number of steps. That any state E_j can be reached from any other state E_i means that $p_{ij}^{(n)} > 0$ for some integer n. This is also referred to as **communicating states**.

As defined in Section 4.4, a matrix $A = [a_{ij}]$ is said to be **positive** if $a_{ij} > 0$ for all i, j. A Markov chain with transition matrix T is said to be **regular** (to avoid unnecessary repetition of the term positive) if there exists an integer N such that T^N is positive.

A regular chain is obviously irreducible. However, the converse is not necessarily true, as can be seen from the simple two-state chain with transition matrix

$$T = \begin{bmatrix} 0 & 1 \\ 1 & 0 \end{bmatrix},$$

since

$$T^{2n} = \begin{bmatrix} 1 & 0 \\ 0 & 1 \end{bmatrix} = I_2 \quad \text{and} \quad T^{2n+1} = \begin{bmatrix} 0 & 1 \\ 1 & 0 \end{bmatrix} = T,$$

for $n = 1, 2, 3 \ldots$. No power of T is a positive matrix.

Example 4.13 *Show that the three-state chain with transition matrix*

$$T = \begin{bmatrix} \frac{1}{3} & \frac{1}{3} & \frac{1}{3} \\ 0 & 0 & 1 \\ 1 & 0 & 0 \end{bmatrix}$$

defines a regular (and hence irreducible) chain.

For the transition matrix T

$$T^2 = \begin{bmatrix} \frac{4}{9} & \frac{1}{9} & \frac{4}{9} \\ 1 & 0 & 0 \\ \frac{1}{3} & \frac{1}{3} & \frac{1}{3} \end{bmatrix}, \qquad T^3 = \begin{bmatrix} \frac{16}{27} & \frac{4}{27} & \frac{7}{27} \\ \frac{1}{3} & \frac{1}{3} & \frac{1}{3} \\ \frac{4}{9} & \frac{1}{9} & \frac{4}{9} \end{bmatrix}.$$

Hence T^3 is a positive matrix, which means that the chain is regular.

An important feature of an irreducible chain is that all its states are of the same type, that is, either all transient or all persistent (null or nonnull), or all have the same period. A proof of this is given by Feller (1968, p. 391). This means that the classification of all states in an irreducible chain can be inferred from the known classification of one state. It is intuitively reasonable to also infer that the states of a finite irreducible chain cannot all be transient, since it would mean that a return to *any* state would not be certain, even though all states are accessible from all other states in a finite number of steps. This requires a proof which will not be included here.

(b) Closed sets

A Markov chain may contain some states which are transient, some which are persistent, absorbing states, and so on. The persistent states can be part of closed subchains. A set of states C in a Markov chain is said to be **closed** if any state within C can be reached from any other state within C, and no state outside C can be reached from any state inside C. Algebraically, a *necessary* condition for this to be the case is that

$$p_{ij} = 0 \quad \forall \quad E_i \in C \quad \text{and} \quad \forall \quad E_j \notin C. \qquad (4.34)$$

An absorbing state is closed with just one state. Note also that a closed subset is itself an irreducible subchain of the full Markov chain.

Example 4.14 *Discuss the status of each state in the six-state Markov chain with transition matrix*

$$T = \begin{bmatrix} \frac{1}{2} & \frac{1}{2} & 0 & 0 & 0 & 0 \\ \frac{1}{4} & \frac{3}{4} & 0 & 0 & 0 & 0 \\ \frac{1}{4} & \frac{1}{4} & \frac{1}{4} & \frac{1}{4} & 0 & 0 \\ \frac{1}{4} & 0 & \frac{1}{4} & \frac{1}{4} & 0 & \frac{1}{4} \\ 0 & 0 & 0 & 0 & \frac{1}{2} & \frac{1}{2} \\ 0 & 0 & 0 & 0 & \frac{1}{2} & \frac{1}{2} \end{bmatrix}.$$

A diagram representing the chain is shown in Figure 4.7. As usual the figure is a great help in settling questions of which sets of states are closed. It can be seen that $\{E_1, E_2\}$ is a closed irreducible subchain since no states outside the states can be reached from E_1 and E_2. Similarly $\{E_5, E_6\}$ is a closed irreducible subchain. The states E_3 and E_4 are transient. All states are aperiodic, which means that E_1, E_2, E_5, and E_6 are ergodic.

(c) Ergodic chains

As we have seen, all the states in an irreducible chain belong to the same class. If all states are ergodic, that is, persistent, nonnull, and aperiodic, then the chain is described as an **ergodic chain**.

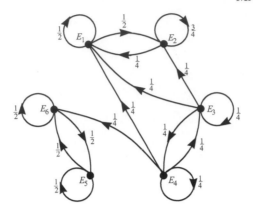

Figure 4.7 *The transition diagram for Example 4.14.*

Example 4.15 *Show that all states of the chain with transition matrix*

$$T = \begin{bmatrix} \frac{1}{3} & \frac{1}{3} & \frac{1}{3} \\ 0 & 0 & 1 \\ 1 & 0 & 0 \end{bmatrix}$$

are ergodic.

This is the same chain as in Example 4.13, where it was shown to be irreducible and regular, which means that all states must be persistent, nonnull, and aperiodic. Hence all states are ergodic.

Example 4.16 *Consider the three-state Markov chain with transition matrix*

$$T = \begin{bmatrix} \frac{1}{5} & \frac{4}{5} & 0 \\ 0 & 0 & 1 \\ 1 & 0 & 0 \end{bmatrix}.$$

Show that all states are ergodic. Find the eigenvalues of T and $Q = \lim_{n\to\infty} T^n$. Determine the mean recurrence times μ_1, μ_2, μ_3 for each state, and confirm that the rows of Q all have the elements $1/\mu_1$, $1/\mu_2$, $1/\mu_3$.

It is easy to check that T^4 is a positive matrix, which implies that the chain is ergodic. The eigenvalues of T are given by

$$\det(T - \lambda I_3) = \begin{vmatrix} \frac{1}{5}-\lambda & \frac{4}{5} & 0 \\ 0 & -\lambda & 1 \\ 1 & 0 & -\lambda \end{vmatrix} = -\lambda^3 + \tfrac{1}{5}\lambda^2 + \tfrac{4}{5}$$

$$= \tfrac{1}{5}(1-\lambda)(5\lambda^2 + 4\lambda + 4) = 0.$$

Hence the eigenvalues can be denoted by

$$\lambda_1 = 1, \qquad \lambda_2 = -\tfrac{2}{5} + \tfrac{4}{5}i, \qquad \lambda_3 = -\tfrac{2}{5} - \tfrac{4}{5}i.$$

The corresponding eigenvectors are

$$\mathbf{r}_1 = \begin{bmatrix} 1 \\ 1 \\ 1 \end{bmatrix}, \qquad \mathbf{r}_2 = \begin{bmatrix} -\tfrac{2}{5} + \tfrac{4}{5}i \\ -\tfrac{1}{2} - i \\ 1 \end{bmatrix}, \qquad \mathbf{r}_3 = \begin{bmatrix} -\tfrac{2}{5} - \tfrac{4}{5}i \\ -\tfrac{1}{2} + i \\ 1 \end{bmatrix}.$$

We use the method of Section 4.4 to find T^n. Let

$$C = \begin{bmatrix} \mathbf{r}_1 & \mathbf{r}_2 & \mathbf{r}_3 \end{bmatrix} = \begin{bmatrix} 1 & -\frac{2}{5}+\frac{4}{5}i & -\frac{2}{5}-\frac{4}{5}i \\ 1 & -\frac{1}{2}-i & -\frac{1}{2}+i \\ 1 & 1 & 1 \end{bmatrix}.$$

The computed inverse is given by

$$C^{-1} = \frac{1}{52} \begin{bmatrix} 20 & 16 & 16 \\ -10-30i & -8+14i & 18+i \\ -10+30i & -8-14i & 18-i \end{bmatrix}.$$

As in Example 4.6, it follows similarly that

$$T^n = C \begin{bmatrix} 1 & 0 & 0 \\ 0 & (-\frac{2}{5}+\frac{4}{5}i)^n & 0 \\ 0 & 0 & (-\frac{2}{5}-\frac{4}{5}i)^n \end{bmatrix} C^{-1}.$$

Hence,

$$Q = \lim_{n \to \infty} T^n = C \begin{bmatrix} 1 & 0 & 0 \\ 0 & 0 & 0 \\ 0 & 0 & 0 \end{bmatrix} C^{-1} = \frac{1}{13} \begin{bmatrix} 5 & 4 & 4 \\ 5 & 4 & 4 \\ 5 & 4 & 4 \end{bmatrix}.$$

The invariant distribution is therefore $\mathbf{p} = \begin{bmatrix} \frac{5}{13} & \frac{4}{13} & \frac{4}{13} \end{bmatrix}$. Note that the elements in $\mathbf{p}$ are the same as the *first row* in C^{-1}. Is this always the case for ergodic chains?

The first returns $f_i^{(n)}$ for each of the states can be easily calculated from the transition diagram Figure 4.8. Thus,

$$f_1^{(1)} = \tfrac{1}{5}, \qquad f_1^{(2)} = 0, \qquad f_1^{(3)} = \tfrac{4}{5}.1.1 = \tfrac{4}{5},$$

$$f_2^{(1)} = f_3^{(1)} = 0, \qquad f_2^{(2)} = f_3^{(2)} = 0, \qquad f_2^{(n)} = f_3^{(n)} = \tfrac{4}{5}\left(\tfrac{1}{5}\right)^{n-3}, \qquad (n \geq 3).$$

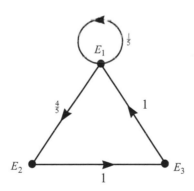

Figure 4.8 *The transition diagram for Example 4.16.*

Hence, by Eqn (4.30),

$$\mu_1 = \sum_{n=1}^{\infty} n f_1^{(n)} = \tfrac{1}{5} + 3.\tfrac{4}{5} = \tfrac{13}{5}.$$

$$\mu_2 = \mu_3 = \sum_{n=1}^{\infty} n f_2^{(n)} = \tfrac{4}{5}\sum_{n=3}^{\infty} n(\tfrac{1}{5})^{n-3} = \tfrac{13}{4}.$$

The vector of reciprocals

$$\begin{bmatrix} \frac{1}{\mu_1} & \frac{1}{\mu_2} & \frac{1}{\mu_3} \end{bmatrix} = \begin{bmatrix} \frac{5}{13} & \frac{4}{13} & \frac{4}{13} \end{bmatrix}$$

agrees with the vector **p** above calculated by the eigenvalue method.

For ergodic chains this is always the case: the *invariant distribution is the vector of mean recurrence time reciprocals*, but we shall not give a proof here.

4.8 A wildlife Markov chain model

Consider the progress of a disease in a closed wildlife population in which births are excluded. This could be, for example, a bird population not in the breeding season. We model the progress of the disease as a discrete Markov chain over time steps given by $0, t, 2t, 3t, \ldots$ which may cover various epochs for t (days, weeks, months, etc.). We start with a population which is susceptible to the disease. The individuals may become infected, from which they either die or recover and become immune: in this model individual birds cannot be re-infected.

The original susceptible population may also die due to other causes. Therefore the population can be divided into four states: E_1 of susceptibles; E_2 infected; E_3 immune; E_4 dead. The proposed transition matrix of the model is

$$T = \begin{bmatrix} p_{11} & p_{12} & 0 & p_{14} \\ 0 & 0 & p_{23} & p_{24} \\ 0 & 0 & p_{33} & p_{34} \\ 0 & 0 & 0 & 1 \end{bmatrix}. \tag{4.35}$$

The transition diagram is shown in Figure 4.9. It is assumed that within any time interval of length t, an infected bird either becomes immune or dies.

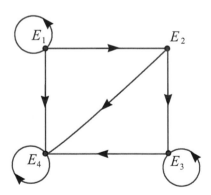

Figure 4.9 *The transition diagram for wild life model.*

The entries in Figure 4.9 are easily explained: p_{11} is the probability that a susceptible is *not* infected, p_{12} that it is infected, p_{14} that a susceptible dies, and so on. Multiple events in any time interval are assumed to be negligible.

If we start with only a susceptible population initially, then we choose

$$\mathbf{p}^{(0)} = \begin{bmatrix} p_1^{(0)} & p_2^{(0)} & p_3^{(0)} & p_4^{(0)} \end{bmatrix} = \begin{bmatrix} 1 & 0 & 0 & 0 \end{bmatrix}.$$

At time nt the probability that a susceptible is in one of the four states is given by

$$\mathbf{p}^{(n)} = \begin{bmatrix} p_1^{(n)} & p_2^{(n)} & p_3^{(n)} & p_4^{(n)} \end{bmatrix} = \mathbf{p}^{(0)} T^n.$$

We could use the eigenvalue method of Section 4.4 to obtain a general solution for T^n. The eigenvalues of T are easy to list, namely $\{p_{11}, 0, p_{33}, 1\}$, since T is an upper triangular matrix. However, even in this case, the eigenvectors are complicated. It is easier to compute powers of T directly.

Some plausible data are applied to the model. An actual wildlife application to house finches in the USA using Markov chains can be found in the paper by Ziplin *et al* (2010) and references cited therein. Here we assume the following representative data for a bird population with a time step $t = 1$ week:

$$
\begin{array}{llll}
p_{11} = 0.79 & p_{12} = 0.20 & p_{13} = 0 & p_{14} = 0.01 \\
p_{21} = 0 & p_{22} = 0 & p_{23} = 0.90 & p_{24} = 0.10 \\
p_{31} = 0 & p_{32} = 0 & p_{33} = 0.99 & p_{34} = 0.01 \\
p_{41} = 0 & p_{42} = 0 & p_{43} = 0 & p_{44} = 1
\end{array}
$$

Hence,

$$T = \begin{bmatrix} 0.79 & 0.20 & 0 & 0.01 \\ 0 & 0 & 0.90 & 0.10 \\ 0 & 0 & 0.99 & 0.01 \\ 0 & 0 & 0 & 1 \end{bmatrix}.$$

We expect from the effect of the absorbing state E_4 (and also from the transition diagram) that

$$T^n \longrightarrow \begin{bmatrix} 0 & 0 & 0 & 1 \\ 0 & 0 & 0 & 1 \\ 0 & 0 & 0 & 1 \\ 0 & 0 & 0 & 1 \end{bmatrix}$$

as $n \to \infty$. To cite a particular case, after 10 weeks,

$$\mathbf{p}^{(10)} = \mathbf{p}^{(0)} T^{10} = \begin{bmatrix} 1 & 0 & 0 & 0 \end{bmatrix} \begin{bmatrix} 0.09 & 0.02 & 0.71 & 0.17 \\ 0 & 0 & 0.82 & 0.18 \\ 0 & 0 & 0.90 & 0.10 \\ 0 & 0 & 0 & 1 \end{bmatrix}$$

$$= \begin{bmatrix} 0.09 & 0.02 & 0.71 & 0.17 \end{bmatrix}$$

(aside from rounding errors).

The probability that a susceptible becomes infected in the time interval $(m-1)t$ to mt is $p_{11}^{m-1} p_{12}$, that is, susceptible for $m-1$ weeks followed by the probability of infection in the following week. Since immunity occurs (that is, no return to susceptibility), then the expected duration d_{12} for a susceptible to become infected is given by (the upper limit is not achievable)

$$d_{12} = \sum_{m=1}^{\infty} m p_{11}^{m-1} p_{12} = \frac{p_{12}}{(1 - p_{11})^2} = 4.53 \text{ weeks.}$$

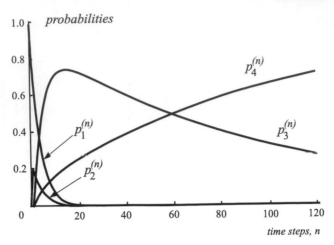

Figure 4.10 *The figure shows list plots of the probabilities* $p_1^{(n)}$ $p_2^{(n)}$ $p_3^{(n)}$ $p_4^{(n)}$ *against the steps* n *as given by* $\mathbf{p}^{(n)} = \mathbf{p}^{(0)} T^n$. *The corresponding states are* E_1 *of susceptibles;* E_2 *infected;* E_3 *immune;* E_4 *dead.*

Figure 4.10 shows how the individual probabilities $p_1^{(n)}$, $p_2^{(n)}$ $p_3^{(n)}$ $p_4^{(n)}$ develop with the time steps.

4.9 Problems

4.1. If $T = [p_{ij}]$, $(i, j = 1, 2, 3)$ and

$$p_{ij} = \frac{i+j}{6+3i},$$

show that T is a row-stochastic matrix. What is the probability that a transition between states E_2 and E_3 occurs at any step?

If the initial probability distribution in a Markov chain is

$$\mathbf{p}^{(0)} = \begin{bmatrix} \frac{1}{2} & \frac{1}{4} & \frac{1}{4} \end{bmatrix},$$

what are the probabilities that states E_1, E_2, and E_3 are occupied after one step? Explain why the probability that the chain finishes in state E_2 is $\frac{1}{3}$ irrespective of the number of steps.

4.2. If

$$T = \begin{bmatrix} \frac{1}{2} & \frac{1}{4} & \frac{1}{4} \\ \frac{1}{3} & \frac{1}{3} & \frac{1}{3} \\ \frac{1}{4} & \frac{1}{2} & \frac{1}{4} \end{bmatrix},$$

calculate $p_{22}^{(2)}$, $p_{31}^{(2)}$, and $p_{13}^{(2)}$.

4.3. For the transition matrix

$$T = \begin{bmatrix} \frac{1}{3} & \frac{2}{3} \\ \frac{1}{4} & \frac{3}{4} \end{bmatrix}$$

calculate $p_{12}^{(3)}$, $p_2^{(2)}$ and $\mathbf{p}^{(3)}$ given that $\mathbf{p}^{(0)} = \begin{bmatrix} \frac{1}{2} & \frac{1}{2} \end{bmatrix}$. Also find the eigenvalues of T, construct a formula for T^n, and obtain $\lim_{n \to \infty} T^n$.

4.4. Sketch transition diagrams for each of the following three-state Markov chains.

(a) $A = \begin{bmatrix} \frac{1}{3} & \frac{1}{3} & \frac{1}{3} \\ 0 & 0 & 1 \\ 1 & 0 & 0 \end{bmatrix}$; (b) $B = \begin{bmatrix} \frac{1}{2} & \frac{1}{4} & \frac{1}{4} \\ 0 & 1 & 0 \\ \frac{1}{2} & \frac{1}{2} & 0 \end{bmatrix}$; (c) $C = \begin{bmatrix} 0 & \frac{1}{2} & \frac{1}{2} \\ 1 & 0 & 0 \\ \frac{1}{3} & \frac{1}{3} & \frac{1}{3} \end{bmatrix}$.

4.5. Find the eigenvalues of the stochastic matrix

$$T = \begin{bmatrix} a & b & c \\ c & a & b \\ b & c & a \end{bmatrix}, \quad (a > 0, b > 0, c > 0 \text{ and } a + b + c = 1).$$

Show that the eigenvalues are complex if $b \neq c$. T is an example of a **doubly stochastic matrix**: the elements in each row *and* column sum to 1.

Find the eigenvalues and eigenvectors in the following cases:
(a) $a = \frac{1}{2}, b = \frac{1}{4}, c = \frac{1}{4}$;
(b) $a = \frac{1}{2}, b = \frac{1}{8}, c = \frac{3}{8}$.

4.6. Find the eigenvalues, eigenvectors, the matrix of eigenvectors C, its inverse C^{-1}, a formula for T^n, and $\lim_{n \to \infty} T^n$ for each of the following transition matrices:
(a)

$$T = \begin{bmatrix} \frac{1}{8} & \frac{7}{8} \\ \frac{1}{2} & \frac{1}{2} \end{bmatrix};$$

(b)

$$T = \begin{bmatrix} \frac{1}{2} & \frac{1}{8} & \frac{3}{8} \\ \frac{1}{4} & \frac{3}{9} & \frac{3}{9} \\ \frac{1}{4} & \frac{5}{8} & \frac{1}{8} \end{bmatrix}.$$

4.7. The weather in a certain region can be characterized as being sunny (S), cloudy (C), or rainy (R) on any particular day. The probability of any type of weather on one day depends only on the state of the weather on the previous day. For example, if it is sunny one day then sun or clouds are equally likely on the next day with no possibility of rain. Explain what other day-to-day possibilities are if the weather is represented by the transition matrix

$T =$		S	C	R
	S	$\frac{1}{2}$	$\frac{1}{2}$	0
	C	$\frac{1}{2}$	$\frac{1}{4}$	$\frac{1}{4}$
	R	0	$\frac{1}{2}$	$\frac{1}{2}$

Find the eigenvalues of T and a formula for T^n. In the long run what percentage of the days are sunny, cloudy, and rainy?

4.8. The eigenvalue method of Section 4.4 for finding general powers of stochastic matrices is only guaranteed to work if the eigenvalues are distinct. Several possibilities occur if the stochastic matrix of a Markov chain has a repeated eigenvalue. The following three examples illustrate these possibilities.

(a) Let

$$T = \begin{bmatrix} \frac{1}{4} & \frac{1}{4} & \frac{1}{2} \\ 1 & 0 & 0 \\ \frac{1}{2} & \frac{1}{4} & \frac{1}{4} \end{bmatrix}$$

be the transition matrix of a three-state Markov chain. Show that T has the repeated eigenvalue $\lambda_1 = \lambda_2 = -\frac{1}{4}$ and $\lambda_3 = 1$, and two distinct eigenvectors

$$\mathbf{r}_1 = \begin{bmatrix} 1 \\ -4 \\ 1 \end{bmatrix} \qquad \mathbf{r}_3 = \begin{bmatrix} 1 \\ 1 \\ 1 \end{bmatrix}.$$

In this case diagonalization of T is not possible. However, it is possible to find a nonsingular matrix C such that

$$T = CJC^{-1},$$

where J is the **Jordan decomposition matrix**[9] given by

$$J = \begin{bmatrix} \lambda_1 & 1 & 0 \\ 0 & \lambda_1 & 0 \\ 0 & 0 & 1 \end{bmatrix} = \begin{bmatrix} -\frac{1}{4} & 1 & 0 \\ 0 & -\frac{1}{4} & 0 \\ 0 & 0 & 1 \end{bmatrix},$$

$$C = \begin{bmatrix} \mathbf{r}_1 & \mathbf{r}_2 & \mathbf{r}_3 \end{bmatrix},$$

and $\mathbf{r}_2$ satisfies

$$(T - \lambda_1 I_3)\mathbf{r}_2 = \mathbf{r}_1.$$

Show that we can choose

$$\mathbf{r}_2 = \begin{bmatrix} -10 \\ 24 \\ 0 \end{bmatrix}.$$

Find a formula for J^n and confirm that, as $n \to \infty$,

$$T^n \to \begin{bmatrix} \frac{12}{25} & \frac{1}{5} & \frac{8}{25} \\ \frac{12}{25} & \frac{1}{5} & \frac{8}{25} \\ \frac{12}{25} & \frac{1}{5} & \frac{8}{25} \end{bmatrix}.$$

(b) A four-state Markov chain has the transition matrix

$$S = \begin{bmatrix} 1 & 0 & 0 & 0 \\ \frac{3}{4} & 0 & \frac{1}{4} & 0 \\ 0 & \frac{1}{4} & 0 & \frac{3}{4} \\ 0 & 0 & 0 & 1 \end{bmatrix}.$$

Sketch the transition diagram for the chain, and note that the chain has two absorbing states and is therefore not a regular chain. Show that the eigenvalues of S are $-\frac{1}{4}, \frac{1}{4}$, and 1 repeated. Show that there are four distinct eigenvectors. Choose the diagonalizing matrix C as

$$C = \begin{bmatrix} 0 & 0 & -4 & 5 \\ -1 & 1 & -3 & 4 \\ 1 & 1 & 0 & 1 \\ 0 & 0 & 1 & 0 \end{bmatrix}.$$

[9] Camille Jordan (1838–1922), French mathematician.

Find its inverse, and show that, as $n \to \infty$,

$$S^n \to \begin{bmatrix} 1 & 0 & 0 & 0 \\ \frac{4}{5} & 0 & 0 & \frac{1}{5} \\ \frac{1}{5} & 0 & 0 & \frac{4}{5} \\ 0 & 0 & 0 & 1 \end{bmatrix}.$$

Note that since the rows are not the same this chain does not have an invariant distribution: this is caused by the presence of two absorbing states.

(c) Show that the transition matrix

$$U = \begin{bmatrix} \frac{1}{2} & 0 & \frac{1}{2} \\ \frac{1}{6} & \frac{1}{3} & \frac{1}{2} \\ \frac{1}{6} & 0 & \frac{5}{6} \end{bmatrix}$$

has a repeated eigenvalue, but that, in this case, three independent eigenvectors can be associated with U. Find a diagonalizing matrix C, and find a formula for U^n using $U^n = CD^n C^{-1}$, where

$$D = \begin{bmatrix} \frac{1}{3} & 0 & 0 \\ 0 & \frac{1}{3} & 0 \\ 0 & 0 & 1 \end{bmatrix}.$$

Confirm also that this chain has an invariant distribution.

4.9. Miscellaneous problems on transition matrices. In each case find the eigenvalues of T, a formula for T^n, and the limit of T^n as $n \to \infty$. The special cases discussed in Problem 4.8 can occur.

(a)
$$T = \begin{bmatrix} \frac{1}{2} & \frac{7}{32} & \frac{9}{32} \\ 1 & 0 & 0 \\ \frac{1}{2} & \frac{1}{4} & \frac{1}{4} \end{bmatrix};$$

(b)
$$T = \begin{bmatrix} \frac{1}{9} & \frac{1}{4} & \frac{5}{12} \\ 1 & 0 & 0 \\ \frac{1}{4} & \frac{1}{4} & \frac{1}{2} \end{bmatrix};$$

(c)
$$T = \begin{bmatrix} \frac{1}{4} & \frac{3}{16} & \frac{9}{16} \\ \frac{3}{4} & 0 & \frac{1}{4} \\ \frac{1}{4} & \frac{1}{4} & \frac{1}{2} \end{bmatrix};$$

(d)
$$T = \begin{bmatrix} \frac{1}{4} & \frac{1}{4} & \frac{1}{2} \\ \frac{5}{12} & \frac{1}{3} & \frac{1}{4} \\ \frac{1}{2} & \frac{1}{4} & \frac{1}{4} \end{bmatrix};$$

(e)
$$T = \begin{bmatrix} 1 & 0 & 0 & 0 \\ \frac{1}{2} & 0 & 0 & \frac{1}{2} \\ 0 & 0 & 1 & 0 \\ 0 & \frac{1}{2} & \frac{1}{2} & 0 \end{bmatrix}.$$

4.10. A four-state Markov chain has the transition matrix

$$T = \begin{bmatrix} \frac{1}{2} & \frac{1}{2} & 0 & 0 \\ 1 & 0 & 0 & 0 \\ \frac{1}{4} & \frac{1}{2} & 0 & \frac{1}{4} \\ \frac{3}{4} & 0 & \frac{1}{4} & 0 \end{bmatrix}.$$

Find f_i, the probability that the chain returns to state E_i, for each state. Determine which states are transient and which are persistent. Which states form a closed subset? Find the eigenvalues of T, and the limiting behavior of T^n as $n \to \infty$.

4.11. A six-state Markov chain has the transition matrix

$$T = \begin{bmatrix} \frac{1}{4} & \frac{1}{2} & 0 & 0 & 0 & \frac{1}{4} \\ 0 & 0 & 0 & 0 & 0 & 1 \\ 0 & \frac{1}{4} & 0 & \frac{1}{4} & \frac{1}{2} & 0 \\ 0 & 0 & 0 & 0 & 1 & 0 \\ 0 & 0 & 0 & \frac{1}{2} & \frac{1}{2} & 0 \\ 0 & 0 & 1 & 0 & 0 & 0 \end{bmatrix}.$$

Sketch its transition diagram. From the diagram, which states do you think are transient and which do you think are persistent? Which states form a closed subset? Determine the invariant distribution in the subset.

4.12. Draw the transition diagram for the seven-state Markov chain with transition matrix

$$T = \begin{bmatrix} 0 & 1 & 0 & 0 & 0 & 0 & 0 \\ 0 & 0 & 1 & 0 & 0 & 0 & 0 \\ \frac{1}{2} & 0 & 0 & \frac{1}{2} & 0 & 0 & 0 \\ 0 & 0 & 0 & 0 & 1 & 0 & 0 \\ 0 & 0 & 0 & 0 & 0 & 1 & 0 \\ \frac{1}{2} & 0 & 0 & 0 & 0 & 0 & \frac{1}{2} \\ 0 & 0 & 0 & 0 & 0 & 0 & 1 \end{bmatrix}.$$

Now discuss the periodicity of the states of the chain. From the transition diagram calculate $p_{11}^{(n)}$ and $p_{44}^{(n)}$ for $n = 2, 3, 4, 5, 6$. (In this example you should confirm that $p_{11}^{(3)} = \frac{1}{2}$ but that $p_{44}^{(3)} = 0$; however, $p_{44}^{(3n)} \neq 0$ for $n = 2, 3, \ldots$ confirming that state E_4 is periodic with period 3.)

4.13. The transition matrix of a three-state Markov chain is given by

$$T = \begin{bmatrix} 0 & \frac{3}{4} & \frac{1}{4} \\ \frac{1}{2} & 0 & \frac{1}{2} \\ \frac{3}{4} & \frac{1}{4} & 0 \end{bmatrix}.$$

Show that $S = T^2$ is the transition matrix of a regular chain. Find its eigenvectors and confirm that S has an invariant distribution given by $\begin{bmatrix} \frac{14}{37} & \frac{13}{37} & \frac{10}{37} \end{bmatrix}$.

4.14. An insect is placed in the maze of cells shown in Figure 4.11. The state E_j is the state in which the insect is in cell j. A transition occurs when the insect moves from one cell to another. Assuming that exits are equally likely to be chosen where there is a choice, construct the transition matrix T for the Markov chain representing the movements of the insect. Show that all states are periodic with period 2. Show that T^2 has two subchains which are both regular. Find the invariant distributions of both subchains. Interpret the results.

4.15. The transition matrix of a four-state Markov chain is given by

$$T = \begin{bmatrix} 1-a & a & 0 & 0 \\ 1-b & 0 & b & 0 \\ 1-c & 0 & 0 & c \\ 1 & 0 & 0 & 0 \end{bmatrix}, \quad (0 < a, b, c < 1).$$

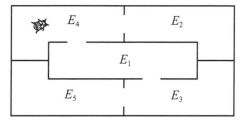

Figure 4.11 *Insect in a maze.*

Draw a transition diagram, and, from the diagram, calculate $f_1^{(n)}$, $(n = 1, 2, \ldots)$, the probability that a first return to state E_1 occurs at the n-th step. Calculate also the mean recurrence time μ_1. What type of state is E_1?

4.16. Show that the transition matrix

$$
T = \begin{bmatrix}
1 - a & a & 0 & 0 \\
1 - a & 0 & a & 0 \\
1 - a & 0 & 0 & a \\
1 & 0 & 0 & 0
\end{bmatrix}
$$

has two imaginary (conjugate) eigenvalues where $0 < a < 1$. If $a = \frac{1}{2}$, confirm that T has the invariant distribution $\mathbf{p} = \begin{bmatrix} \frac{8}{15} & \frac{4}{15} & \frac{2}{15} & \frac{1}{15} \end{bmatrix}$.

4.17. A production line consists of two manufacturing stages. At the end of each manufacturing stage each item in the line is inspected, where there is a probability p that it will be scrapped, q that it will be sent back to that stage for reworking, and $(1 - p - q)$ that it will be passed to the next stage or completed. The production line can be modelled by a Markov chain with four states: E_1, item scrapped; E_2, item completed; E_3, item in first manufacturing stage; E_4, item in second manufacturing stage. We define states E_1 and E_2 to be absorbing states so that the transition matrix of the chain is

$$
T = \begin{bmatrix}
1 & 0 & 0 & 0 \\
0 & 1 & 0 & 0 \\
p & 0 & q & 1 - p - q \\
p & 1 - p - q & 0 & q
\end{bmatrix}.
$$

An item starts along the production line. What is the probability that it is completed in two stages? Calculate $f_3^{(n)}$ and $f_4^{(n)}$. Assuming that $0 < p + q < 1$, what kind of states are E_3 and E_4? What is the probability that an item starting along the production line is ultimately completed?

4.18. The step-dependent transition matrix of Example 4.10 is

$$
T_n = \begin{bmatrix}
\frac{1}{2} & \frac{1}{2} & 0 \\
0 & 0 & 1 \\
1/(n+1) & 0 & n/(n+1)
\end{bmatrix}, \qquad (n = 1, 2, 3, \ldots).
$$

Find the mean recurrence time for state E_3, and confirm that E_3 is a persistent, nonnull state.

4.19. In Example 4.10, a persistent, null state occurred in a chain with step-dependent transitions: such a state cannot occur in a finite chain with a constant transition matrix. However, chains over an infinite number of states can have persistent, null states. Consider the following chain which has an infinite number of states $E_1, E_2, \ldots$ with the transition probabilities

$$p_{11} = \frac{1}{2}, \quad p_{12} = \frac{1}{2}, \quad p_{j1} = \frac{1}{j+1}, \quad p_{j,j+1} = \frac{j}{j+1}, \quad (j \geq 2).$$

Find the mean recurrence time for E_1, and confirm that E_1 is a persistent, null state.

4.20. A random walk takes place on $1, 2, \ldots$ subject to the following rules. A jump from position i to position 1 occurs with probability q_i, and from position i to $i+1$ with probability $1 - q_i$ for $i = 1, 2, 3, \ldots$, where $0 < q_i < 1$. Sketch the transition diagram for the chain. Explain why to investigate the persistence of every state, only one state, say state 1, need be considered. Show that the probability that a first return to state 1 occurs at some step is

$$f_1 = \sum_{j=1}^{\infty} \left[\prod_{k=1}^{j-1} (1 - q_k) \right] q_j.$$

If $q_j = q \ (j = 1, 2, \ldots)$, show that every state is persistent.

4.21. *A Markov chain maze.* Figure 4.12 shows a maze with entrance E_1 and further gates E_2, E_3, E_4, E_5, and E_6 with target E_7. These gates can be represented by states in a Markov

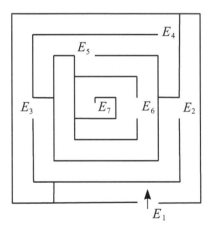

Figure 4.12 *The maze in Problem 4.21.*

chain. At each $E_1, \ldots, E_6$ there are two possible new paths which are assumed equally likely to be chosen. The target E_7 can be considered to be an absorbing state. Construct a 7×7 transition matrix for the maze assuming that the walker does not return to a previous state and does not learn from previous choices: for example, at E_1 he or she can still make the mistake of walking the dead-end again. Find the probabilities that the walker reaches the target in $6, 7, 8, \ldots$ steps.

Suppose now that the walker learns from wrong choices. To accommodate this, let E_{11} be the entrance and E_{12} the return to the entrance after a wrong choice (the dead-end); let E_{21} and E_{22} have the same meaning for the second entrance, and so on. Hence the probabilities are:

$$\mathbf{P}(E_{12} \to E_{12}) = \tfrac{1}{2}; \quad \mathbf{P}(E_{11} \to E_{21}) = \tfrac{1}{2}; \quad \mathbf{P}(E_{12} \to E_{21}) = 1;$$

$$\mathbf{P}(E_{21} \to E_{22}) = \tfrac{1}{2}; \quad \mathbf{P}(E_{31} \to E_{21}) = \tfrac{1}{2}; \quad \mathbf{P}(E_{22} \to E_{31}) = 1;$$

and similarly for the remaining probabilities. The transition matrix is now 13×13. with states

$$E_{11}, E_{12}, E_{21}, E_{22}, \dots, E_{61}, E_{62}, E_{7}.$$

Find the probabilities that the walker reaches the centre in 6,7,8 steps. (One really needs a computer program to compute the matrix products.)

4.22. In a finite Markov chain a subset C of states is said to be closed if, for states i and j,

$$i \in C, \quad \text{then transition } i \to j \Rightarrow \quad j \in C.$$

Find the closed subset in a chain with transition matrix

$$T = \begin{bmatrix} 0 & 0 & \frac{1}{2} & \frac{1}{2} & 0 & 0 \\ 0 & 0 & \frac{1}{2} & 0 & \frac{1}{4} & \frac{1}{4} \\ 0 & 0 & \frac{1}{3} & \frac{1}{3} & 0 & \frac{1}{3} \\ 0 & 0 & 0 & 0 & 0 & 1 \\ \frac{1}{2} & 0 & 0 & 0 & 0 & \frac{1}{2} \\ 0 & 0 & 1 & 0 & 0 & 0 \end{bmatrix}.$$

4.23. Chess knight moves. A knight moves on a reduced chess board with $4 \times 4 = 16$ squares. The knight starts at the bottom left-hand corner and moves according to the usual chess rules. Treating moves as a Markov chain, construct a 16×16 transition matrix for the knight moving from any square (easier if you design a computer program for the matrix). Show that the knight returns to it, starting corner after 2,4,6 moves (must be even) with probabilities $\frac{1}{4}, \frac{1}{6}, \frac{7}{54}$, respectively. Find the corresponding first returns. [Just a reminder: if $f_{11}^{(n)}$ is the probability that the first return to corner $(1, 1)$ (say) after n moves, and $p_{11}(m)$ is the probability that the knight is at $(1, 1)$ after m moves], then

$$f_{11} = p_{11}^{(1)}, \quad f_{11}^{(n)} = p_{11}^{(n)} - \sum_{m=1}^{n-1} f_{11}^{(m)} p_{11}^{(n-m)}, \quad (n \geq 2).]$$

CHAPTER 5

Poisson Processes

5.1 Introduction

In many applications of stochastic processes, the random variable can be a continuous function of the time t. For example, in a population, births and deaths can occur at any time, and any random variable representing such a probability model must take account of this dependence on time. Other examples include the arrival of telephone calls at an office, or the emission of radioactive particles recorded on a Geiger counter. Interpreting the term population in the broad sense (not simply humans and animals, but particles, telephone calls, etc., depending on the context), we might be interested typically in the probability that the population size is, say, n at time t. We shall represent this probability usually by $p_n(t)$. For the Geiger[1] counter application it will represent the probability that n particles have been recorded up to time t, whilst for the arrival of telephone calls it could represent the number of calls logged up to time t. These are examples of a process with discrete states but observed over continuous times.

5.2 The Poisson process

Let $N(t)$ be a time-varying random variable representing the population size at time t. Consider the probability of population size n at time t given by

$$p_n(t) = \mathbf{P}[N(t) = n] = \frac{(\lambda t)^n e^{-\lambda t}}{n!}, \qquad (t \geq 0) \tag{5.1}$$

for $n = 0, 1, 2, \ldots$ (remember $0! = 1$). It is assumed that $N(t)$ can take the integer values $n = 0, 1, 2, \ldots$. We can confirm that (5.1) is a probability distribution by observing that

$$\sum_{n=0}^{\infty} p_n(t) = \sum_{n=0}^{\infty} \frac{(\lambda t)^n}{n!} e^{-\lambda t} = e^{-\lambda t} \sum_{n=0}^{\infty} \frac{(\lambda t)^n}{n!} = e^{-\lambda t} e^{\lambda t} = 1,$$

using the power series expansion for the exponential function (see the Appendix). Note that $p_0(0) = 1$, that is, the initial population size is 1.

In fact (see Section 1.7), $p_n(t)$ is a **Poisson probability (mass) function** with **parameter** or **intensity** λ. For this reason any application for which Eqn (5.1) holds

---

[1] Hans Geiger (1882–1945), German physicist.

is known as a **Poisson process**: we shall give a full formal statement of the process
in Section 5.7. Examples of some probabilities for $p_n(t)$ versus λt in the cases $n =
0, 1, 2, 3$ are shown in Figure 5.1. The Poisson process is a special case of the more
general birth process which is developed in Chapter 6.

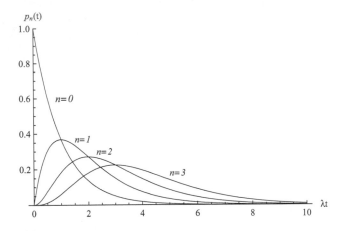

Figure 5.1 *Probabilities $p_n(t)$ for $n = 0, 1, 2, 3$.*

Since, for $n \geq 1$,
$$\frac{dp_n(t)}{dt} = \frac{(n - \lambda t)}{n!} \lambda^n t^{n-1} e^{-\lambda t}, \tag{5.2}$$
the maximum values of the probabilities for fixed n occur at time $t = n/\lambda$, where
$dp_n(t)/dt = 0$. For $n = 0$,
$$\frac{dp_0(t)}{dt} = -\lambda e^{-\lambda t}. \tag{5.3}$$

The mean $\mu(t)$ of the Poisson distribution will be a function of time given by the
sum of the products of the possible outcomes n and their probable occurrences $p_n(t)$,
$(n = 0, 1, 2, \ldots)$, namely,

$$
\begin{aligned}
\mu(t) = \mathbf{E}[N(t)] \ &= \ \sum_{n=0}^{\infty} n p_n(t) = \sum_{n=1}^{\infty} n \frac{(\lambda t)^n}{n!} e^{-\lambda t} \\
&= \ e^{-\lambda t} \lambda t \sum_{n=1}^{\infty} \frac{(\lambda t)^{n-1}}{(n-1)!} = e^{-\lambda t} \lambda t e^{\lambda t} = \lambda t. \quad (5.4)
\end{aligned}
$$

Note that the mean value increases linearly with time at rate λ.

This observation that the mean increases steadily with time gives a pointer as to
why the Poisson distribution is appropriate for cumulative recording, such as the
Geiger counter application and the call-logging problem for the telephone operator.
For the Geiger counter, the recording of radioactive particle hits is assumed to occur
at random with probability of a new subsequent hit being independent of any previous
recordings. This independence requirement is crucial for the Poisson process. We can
show how this arises from the Poisson process as follows.

Example 5.1. *Find the variance of the Poisson distribution $p_n(t)$ $(n = 0, 1, 2, \ldots)$.*

Expressing the variance in terms of means, the variance is given by (see Section 1.6)

$$\mathbf{V}[N(t)] = \mathbf{E}[N(t)^2] - [\mathbf{E}[N(t)]]^2 = \mathbf{E}[N(t)^2] - (\lambda t)^2$$

by (5.4). Also

$$
\begin{aligned}
\mathbf{E}[N(t)^2] &= \sum_{n=1}^{\infty} \frac{n^2 (\lambda t)^n}{n!} e^{-\lambda t} = e^{-\lambda t} \sum_{n=1}^{\infty} \frac{n(\lambda t)^n}{(n-1)!} \\
&= e^{-\lambda t} t \frac{d}{dt} \sum_{n=1}^{\infty} \frac{\lambda^n t^n}{(n-1)!} = e^{-\lambda t} t \frac{d}{dt} [\lambda t e^{\lambda t}] = \lambda t + (\lambda t)^2.
\end{aligned}
$$

Thus

$$\mathbf{V}[N(t)] = \lambda t + (\lambda t)^2 - (\lambda t)^2 = \lambda t.$$

Note that the Poisson distribution has the property that its mean is the same as its variance.

From Eqns (5.1), (5.2), and (5.3) it follows that

$$\frac{dp_0(t)}{dt} = -\lambda p_0(t), \qquad (5.5)$$

$$
\begin{aligned}
\frac{dp_n(t)}{dt} &= \frac{(n - \lambda t)}{n!} \lambda^n t^{n-1} e^{-\lambda t} = \frac{\lambda^n}{(n-1)!} t^{n-1} e^{-\lambda t} - \frac{\lambda^{n+1}}{n!} t^n e^{-\lambda t} \\
&= \lambda[p_{n-1}(t) - p_n(t)], \qquad (n \geq 1). \qquad (5.6)
\end{aligned}
$$

These are **differential-difference equations** for the sequence of probabilities $p_n(t)$.
From the definition of differentiation, the derivatives are obtained by the limiting process

$$\frac{dp_n(t)}{dt} = \lim_{\delta t \to 0} \frac{p_n(t + \delta t) - p_n(t)}{\delta t},$$

so that approximately, for small $\delta t > 0$,

$$\frac{dp_n(t)}{dt} \approx \frac{p_n(t + \delta t) - p_n(t)}{\delta t}.$$

Thus eliminating the derivatives in Eqns (5.5) and (5.6) in favor of their approximations, we can replace the equations by

$$\frac{p_0(t + \delta t) - p_0(t)}{\delta t} \approx -\lambda p_0(t),$$

$$\frac{p_n(t + \delta t) - p_n(t)}{\delta t} \approx \lambda[p_{n-1}(t) - p_n(t)],$$

so that

$$\left.
\begin{aligned}
p_0(t + \delta t) &\approx (1 - \lambda \delta t) p_0(t) \\
p_n(t + \delta t) &\approx p_{n-1}(t) \lambda \delta t + p_n(t)(1 - \lambda \delta t), \qquad (n \geq 1)
\end{aligned}
\right\}. \qquad (5.7)
$$

We can interpret the equations as follows. For the Geiger counter, we can infer from these formulas that the probability that a particle is recorded in the short time interval δt is $\lambda \delta t$, and that the probability that two or more particles are recorded is negligible, and consequently that no recording takes place with probability $(1 - \lambda \delta t)$. In Eqn (5.7) the only way in which the outcome reading $n(\geq 1)$ can occur at time $t + \delta t$ is that either one particle was recorded in the interval δt when n particles were recorded at time t, or that nothing occurred with probability $(1 - \lambda \delta t)$ when $n - 1$ particles were recorded at time t. In fact, Eqn (5.7) is really a re-statement of the **partition theorem** or **law of total probability** (Section 1.3), and is often the starting point for the modeling of random processes in continuous time. We will look at this approach for the Poisson process in the next section, before we develop it further in the next chapter for birth and death processes and queues.

5.3 Partition theorem approach

We can use Eqn (5.7) as an approach using the partition theorem (Section 1.3). We argue as in the last paragraph of the previous section but tighten the argument. For the Geiger counter application (a similar argument can be adapted for the call-logging problem, etc.) we assume that the probability that one particle is recorded in the short time interval δt is

$$\lambda \delta t + o(\delta t).$$

(The term $o(\delta t)$ described in words as 'little o δt' means that the remainder or error is *of lower order than* δt, that is [(see the Appendix)],

$$\lim_{\delta t \to 0} \frac{o(\delta t)}{\delta t} = 0.)$$

The probability of two or more hits is assumed to be $o(\delta t)$, that is, negligible as $\delta t \to 0$, and the probability of no hits is $1 - \lambda \delta t + o(\delta t)$. We now apply the partition theorem on the possible outcomes. The case $n = 0$ is special since reading zero can only occur through no event occurring. Thus

$$p_0(t + \delta t) = [1 - \lambda \delta t + o(\delta t)]p_0(t),$$

$$p_n(t + \delta t) = p_{n-1}(t)(\lambda \delta t + o(\delta t)) + p_n(t)(1 - \lambda \delta t + o(\delta t)) + o(\delta t), \ (n \geq 1).$$

Dividing through by δt and re-organizing the equations, we find that

$$\frac{p_0(t + \delta t) - p_0(t)}{\delta t} = -\lambda p_0(t) + o(1),$$

$$\frac{p_n(t + \delta t) - p_n(t)}{\delta t} = \lambda[p_{n-1}(t) - p_n(t)] + o(1).$$

Now let $\delta t \to 0$. Then, by the definition of the derivative and $o(1)$,

$$\frac{dp_0(t)}{dt} = -\lambda p_0(t), \tag{5.8}$$

$$\frac{dp_n(t)}{dt} = \lambda[p_{n-1}(t) - p_n(t)], \qquad (n \geq 1). \tag{5.9}$$

All terms $o(1)$ tend to zero as $\delta t \to 0$. Not surprisingly, we have recovered Eqns (5.5) and (5.6).

We started this chapter by looking at the Poisson distributions defined by Eqn (5.1). We now look at techniques for solving Eqns (5.8) and (5.9), although for the Poisson process we know the solutions. The methods give some insight into the solution of more general continuous-time random processes.

5.4 Iterative method

Suppose that our model of the Geiger counter is based on Eqn (5.8), and that we wish to solve the equations to recover $p_n(t)$, which we assume to be unknown for the purposes of this exercise. Equation (5.8) is an ordinary differential equation for one unknown function $p_0(t)$. It is of first-order, and it can be easily verified that its general solution is

$$p_0(t) = C_0 e^{-\lambda t}, \tag{5.10}$$

where C_0 is a constant. We need to specify **initial conditions** for the problem. Assume that the instrumentation of the Geiger counter is set to zero initially. Thus we have the *certain* event for which $p_0(0) = 1$, and consequently $p_n(0) = 0$, $(n \geq 1)$: the probability of any reading other than zero is zero at time $t = 0$. Hence $C_0 = 1$ and

$$p_0(t) = e^{-\lambda t}. \tag{5.11}$$

Now put $n = 1$ in (5.9) so that

$$\frac{dp_1(t)}{dt} = \lambda p_0(t) - \lambda p_1(t),$$

or

$$\frac{dp_1(t)}{dt} + \lambda p_1(t) - \lambda p_0(t) = \lambda e^{-\lambda t},$$

after substituting for $p_0(t)$ from Eqn (5.11). This first order differential equation for $p_1(t)$ is of **integrating factor** type with integrating factor

$$e^{\int \lambda dt} = e^{\lambda t},$$

in which case it can be rewritten as the following separable differential equation:

$$\frac{d}{dt}\left(e^{\lambda t}p_1(t)\right) = \lambda e^{\lambda t}e^{-\lambda t} = \lambda.$$

Hence, integration with respect to t results in

$$e^{\lambda t}p_1(t) = \int \lambda dt = \lambda t + C_1.$$

Thus

$$p_1(t) = \lambda t e^{-\lambda t} + C_1 e^{-\lambda t} = \lambda t e^{-\lambda t}$$

since $p_1(0) = 0$. We now repeat the process by putting $n = 2$ in Eqn (5.9) and substituting in the $p_1(t)$ which has just been found. The result is the equation

$$\frac{dp_2(t)}{dt} + \lambda p_2(t) = \lambda p_1(t) = \lambda t e^{-\lambda t}.$$

This is a further first-order integrating-factor differential equation which can be solved using the same method. The result is, using the initial condition $p_2(0) = 0$,

$$p_2(t) = \frac{(\lambda t)^2 e^{-\lambda t}}{2!}.$$

The method can be repeated for $n = 3, 4, \ldots$, and the results imply that

$$p_n(t) = \frac{(\lambda t)^n e^{-\lambda t}}{n!}, \tag{5.12}$$

which is the probability given by Eqn (5.1) as we would expect. The result in Eqn (5.12) can be justified rigorously by constructing a proof by induction.

The initial condition is built into the definition of the Poisson process (see the summary in Section 5.8). The iterative approach outlined above does permit the use of other initial conditions such as the assumption that the Geiger counter is set to reading n_0, say, at time $t = 0$ (see Problem 5.8).

The iterative method works for the differential-difference equations (5.9) because they contain forward differencing only in n. In many applications both forward and backward differencing appear in the equations with the result that successive methods of solution can no longer be applied. The alternative generating function approach will be explained in the next section.

5.5 The generating function

An alternative approach to the solution of differential-difference equations uses the **probability generating function** first introduced in Section 1.9. For continuous-time random processes, we define a generating function as a power series in a dummy variable s, say, in which the coefficients in the series are the probabilities, the $p_n(t)$'s in our notation here. Thus we construct a generating function $G(s, t)$ as

$$G(s, t) = \sum_{n=0}^{\infty} p_n(t) s^n. \tag{5.13}$$

Here s is a dummy variable which is in itself not of much direct interest. However, given the function $G(s, t)$, $p_n(t)$ can be recovered from the series obtained by expanding $G(s, t)$ in powers of s, and looking at their coefficients.

The practical steps involved in expressing the differential-difference equations in terms of the generating function are as follows. Multiply Eqn (5.9) by s^n, and add the equations together including Eqn (5.8). In summation notation the result is

$$\sum_{n=0}^{\infty} \frac{dp_n(t)}{dt} s^n = \lambda \sum_{n=1}^{\infty} p_{n-1}(t) s^n - \lambda \sum_{n=0}^{\infty} p_n(t) s^n. \tag{5.14}$$

Note carefully the lower limits on the summations. Note also that the right-hand side of Eqn (5.8) has been included in the second series on the right-hand side of Eqn (5.14). We attempt to express each of the series in Eqn (5.14) in terms of the generating function $G(s, t)$ or its partial derivatives with respect to either s or t.

Thus looking at each of the series in turn, we find that

$$\sum_{n=0}^{\infty} \frac{dp_n(t)}{dt} s^n = \frac{\partial}{\partial t} \sum_{n=0}^{\infty} p_n(t) s^n = \frac{\partial G(s,t)}{\partial t},$$

$$\sum_{n=1}^{\infty} p_{n-1}(t) s^n = \sum_{m=0}^{\infty} p_m(t) s^{m+1} = sG(s,t), \quad \text{(putting } n = m + 1\text{)},$$

$$\sum_{n=0}^{\infty} p_n(t) s^n = G(s,t).$$

We can now replace Eqn (5.14) by the equation

$$\frac{\partial G(s,t)}{\partial t} = \lambda s G(s,t) - \lambda G(s,t) = \lambda(s-1)G(s,t). \tag{5.15}$$

This *partial* differential equation in $G(s,t)$ replaces the set of differential-difference Eqns (5.8) and (5.9). If Eqn (5.15) can be solved for the generating function $G(s,t)$, and if this function can then be expanded in powers of s, then the probabilities can be read off from this expansion, and it is possible by using the uniqueness property of probability generating functions that the distribution could be identified by name.

Since Eqn (5.15) only contains a derivative with respect to t, it behaves more like an ordinary differential equation. We can integrate it with respect to t (it is a similar equation to (5.8)) so that

$$G(s,t) = A(s)e^{\lambda(s-1)t}, \tag{5.16}$$

where $A(s)$ is the 'constant of integration' but nevertheless can be a function of the other variable s. That (5.16) is the solution can be verified by direct substitution in Eqn (5.15).

The function $A(s)$ is determined by the initial conditions which must now be expressed in terms of the generating function. The initial conditions $p_0(0) = 1$, $p_n(0) = 0$, $(n \geq 1)$ translate into

$$G(s,0) = \sum_{n=0}^{\infty} p_n(0) s^n = 1.$$

Generally the initial conditions will lead to $G(s,0)$ being a specified function in s. Thus in Eqn (5.16), $A(s) = 1$ and the required generating function for this Poisson process is

$$G(s,t) = e^{\lambda(s-1)t}. \tag{5.17}$$

To obtain the individual probabilities, we expand the generating function in powers of s. In this case we need the power series for the exponential function $e^{\lambda st}$. Applying this result to $G(s,t)$, we obtain

$$G(s,t) = e^{\lambda(s-1)t} = e^{-\lambda t}e^{\lambda st} = e^{-\lambda t}\sum_{n=0}^{\infty}\frac{(\lambda st)^n}{n!}$$

$$= \sum_{n=0}^{\infty}\frac{(\lambda t)^n e^{-\lambda t}}{n!}s^n.$$

Hence the probability $p_n(t)$ is confirmed again as

$$p_n(t) = \frac{(\lambda t)^n e^{-\lambda t}}{n!}, \qquad (n = 0, 1, 2, \ldots).$$

The mean value at time t can also be expressed in terms of the generating function. Quite generally the mean is given by (see Section 1.9)

$$\mu(t) = \sum_{n=1}^{\infty} n p_n(t) = \left[\frac{\partial G(s,t)}{\partial s} \right]_{s=1} = G_s(1,t). \qquad (5.18)$$

For the Poisson process above,

$$\mu(t) = \left[\frac{\partial}{\partial s} e^{\lambda(s-1)t} \right]_{s=1} = \lambda t e^{\lambda(s-1)t} \Big|_{s=1} = \lambda t, \qquad (5.19)$$

which confirms (5.4) again.

The variance of the Poisson process can be found using the probability generating function again. Modifying the result for $G(s,t)$, we find, for fixed t, that

$$\sigma^2 = G_{ss}(1,t) + G_s(1,t) - G_s(1,t)^2. \qquad (5.20)$$

Hence by (5.13),

$$\begin{aligned} \sigma^2 &= [\lambda^2 t^2 e^{\lambda(s-1)t} + \lambda t e^{\lambda(s-1)t} - (\lambda t)^2 e^{2\lambda(s-t)}]_{s=1} \\ &= \lambda^2 t^2 + \lambda t - (\lambda t)^2 = \lambda t, \end{aligned} \qquad (5.21)$$

which happens to be the same as the mean.

5.6 Arrival times

For the Geiger counter, the **arrival time** T_n for reading n is defined as the earliest time at which the random variable $N(t) = n$. We can set $T_0 = 0$. Figure 5.2 shows how the times might occur in a Poisson process. The **inter-arrival time** $Q_n = T_n -$

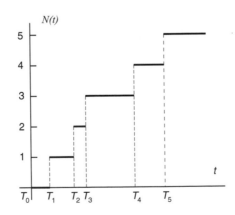

Figure 5.2 *The Poisson process and some typical arrival times* $T_0, T_1, T_2, \ldots$.

T_{n-1} is the time between successive hits. We have assumed that the particles hit the Geiger counter randomly and independently. Hence we might reasonably expect that the random variable giving the arrival time of the first particle will have the same distribution as the inter-arrival times between any two successive readings. It is as if the counter has reset itself to zero and is awaiting the arrival of the next particle, that is, it has the no-memory property (see Problem 5.2). At time t the probability that no particle has been detected by the Geiger counter is $p_0(t) = e^{-\lambda t}$. The distribution function of the arrival times of the *next* reading is therefore given by

$$F(x) = 1 - p_0(x) = 1 - e^{-\lambda x}, \quad (x \geq 0),$$

which leads to the exponential probability density function

$$f(x) = \frac{dF(x)}{dx} = \lambda e^{-\lambda x}, \quad (x \geq 0).$$

The mean or expected value of the inter-arrival times is given by

$$\mu = \int_0^\infty \lambda x e^{-\lambda x} dx \tag{5.22}$$

$$= \left[-x e^{-\lambda x} \right]_0^\infty + \int_0^\infty e^{-\lambda x} dx, \quad \text{(integrating by parts)} \tag{5.23}$$

$$= 0 + \left[-\frac{e^{-\lambda x}}{\lambda} \right]_0^\infty = \frac{1}{\lambda}. \tag{5.24}$$

Hence the mean of the inter-arrival times or the times between successive readings for the Geiger counter is the reciprocal of the rate λ. The variance is given by

$$\sigma^2 = \int_0^\infty (x - \mu)^2 f(x) dx = \int_0^\infty \left(x - \frac{1}{\lambda} \right)^2 \lambda e^{-\lambda x} dx$$

$$= \frac{1}{\lambda^2}.$$

Example 5.2 *Incoming telephone calls to an operator are assumed to be a Poisson process with parameter λ. Find the density function of the length of time for n calls to be received, and find the mean time and variance of the random variable of the length of time for n calls.*

We are now interested in the time T_n, which is the earliest time at which the random variable $N(t) = n$ occurs, and its distribution. The probability distribution of the random variable T_n is given by

$$\begin{aligned} F(t) &= \mathbf{P}\{T_n \leq t\} \\ &= \mathbf{P}\{n \text{ or more calls have arrived in the time interval } (0, t)\} \\ &= p_n(t) + p_{n+1}(t) + \cdots \\ &= \sum_{r=n}^\infty p_r(t) = \sum_{r=n}^\infty \frac{(\lambda t)^r e^{-\lambda t}}{r!} \end{aligned}$$

using (5.1). The corresponding density function is

$$f(t) = \frac{dF(t)}{dt} = \sum_{r=n}^\infty \left[\frac{\lambda(\lambda t)^{r-1} e^{-\lambda t}}{(r-1)!} - \frac{\lambda(\lambda t)^r e^{-\lambda t}}{r!} \right] = \frac{\lambda(\lambda t)^{n-1}}{(n-1)!} e^{-\lambda t},$$

which is the density function of a gamma distribution with parameters n and λ. The mean of this density function is

$$\mu = \mathbf{E}(T_n) = \int_0^\infty tf(t)dt = \int_0^\infty \frac{(\lambda t)^n e^{-\lambda t}}{(n-1)!} dt = \frac{1}{\lambda(n-1)!} \int_0^\infty u^n e^{-u} du = \frac{n}{\lambda}$$

(see Section 1.8).

The variance of this gamma distribution is given by (see Section 1.8 again)

$$
\begin{aligned}
\mathbf{V}(T_n) &= \mathbf{E}(T_n^2) - [\mathbf{E}(T_n)]^2 = \int_0^\infty \frac{t^2 \lambda (\lambda t)^{n-1}}{(n-1)!} e^{-\lambda t} dt - \mu^2 \\
&= \frac{1}{\lambda(n-1)!} \int_0^\infty (\lambda t)^{n+1} e^{-\lambda t} dt - \frac{n^2}{\lambda^2} \\
&= \frac{1}{\lambda^2(n-1)!} \int_0^\infty u^{n+1} e^{-u} du - \frac{n^2}{\lambda^2} \\
&= \frac{(n+1)!}{\lambda^2(n-1)!} - \frac{n^2}{\lambda^2} = \frac{n(n+1)}{\lambda^2} - \frac{n^2}{\lambda^2} = \frac{n}{\lambda^2}.
\end{aligned}
$$

Example 5.3 *A fire and emergency rescue service receives calls for assistance at a rate of ten per day. Teams man the service in twelve hour shifts. Assume that requests for help form a Poisson process.*

(i) From the beginning of a shift, how long would the team expect to wait until their first call?

(ii) What is the probability that a team would receive six requests for help in a shift?

(iii) What is the probability that a team has no requests for assistance in a shift

(iv) Of calls for assistance, one in five is a false alarm. What is the probability that a team has six requests for help in a shift but no false alarms?

Let time be measured in hours. Then 10 calls per day is equivalent to $10/24$ per hour. Hence the Poisson rate is $\lambda = 0.4167$. In this problem, $p_n(t) = (\lambda t)^n e^{-\lambda t}/n!$ is the probability that there are n calls in time t, where t is measured in hours.

(i) From Eqn (5.24) the team would expect to wait for $1/\lambda = 2.4$ hours until the first emergency.

(ii) Using Eqn (5.12), the required probability is

$$\mathbf{P}[N(12) = 6] = p_6(12) = \frac{1}{6!} \left[\frac{10}{24}.12 \right]^6 e^{-\frac{10}{24}.12} = 0.146.$$

(iii) The probability of no request will be

$$\mathbf{P}[N(12) = 0] = p_0(12) = e^{-\frac{10}{24}.12} = e^{-5} = 0.00674.$$

(iv) Then

$$
\begin{aligned}
\mathbf{P}[N = 6, \text{no false alarm}] &= \mathbf{P}[\text{no false alarm}|N = 6]\mathbf{P}[N = 6] \\
&= \left(\frac{4}{5} \right)^6 p_6(12) \\
&= 0.262 \times 0.146 = 0.038
\end{aligned}
$$

5.7 Summary of the Poisson process

We have introduced the Poisson process in an informal way mainly through the illustrative model application of the Geiger counter. We can formally summarize what is meant by a Poisson process as follows.

Consider the random process with the random variable $N(t)$, $(t \geq 0)$. Then $N(t)$ is generated by a Poisson process if

- (i) $N(t)$ can take the values $\{0, 1, 2, \ldots\}$;
- (ii) $N(0) = 0$;
- (iii) $N(t_1) \leq N(t_2)$ if $t_2 \geq t_1$;
- (iv) for any sequence $0 < t_1 < t_2 \cdots < t_n$, the random variables $N(t_i) - N(t_{i-1})$ are mutually independent;
- (v) $\mathbf{P}(N(t + \delta t) = n + 1 | N(t) = n) = \lambda \delta t + o(\delta t)$, $\mathbf{P}(N(t + \delta t) \geq n + 2 | N(t) = n) = o(\delta t)$.
- (vi) the probability generating function is

$$G(s, t) = e^{\lambda(s-1)t},$$

  subject to $p_0(0) = 1$.
- (vii) the inter-arrival times are exponentially distributed.

In (v) the first conditional probability specifies that the probability of an event in the time interval $(t, t + \delta t)$ behaves linearly in δt, whilst the second conditional probability states that the likelihood of two or more events taking place in this time interval is negligible.

5.8 Problems

5.1. The number of cars which pass a roadside speed camera within a specified hour is assumed to be a Poisson process with intensity $\lambda = 92$: on average 92 cars pass in the hour. It is also found that 1% of cars exceed the designated speed limit. What are the probabilities that (a) at least one car exceeds the speed limit, (b) at least two cars exceed the speed limit in the hour?

5.2. If the between-event time in a Poisson process has an exponential distribution with intensity λ and with density $\lambda e^{-\lambda t}$, then the probability that the time T for the next event to occur is at least t is

$$P\{T > t\} = e^{-\lambda t}.$$

Show that, if $t_1, t_2 \geq 0$, then

$$\mathbf{P}\{T > t_1 + t_2 | T > t_1\} = \mathbf{P}\{T > t_2\}.$$

What does this result imply about the Poisson process and its memory of past events?

5.3. The number of cars which pass a roadside speed camera is assumed to behave as a Poisson process with intensity λ. It is found that the probability that a car exceeds the designated speed limit is p.

(a) Show that the number of cars which break the speed limit also forms a Poisson process.

(b) If n cars pass the cameras in time t, find the probability function for the cars which exceed the speed limit.

5.4. The variance of a random variable $X(t)$ is given by

$$\mathbf{V}(X(t)) = \mathbf{E}(X(t)^2) - \mathbf{E}(X(t))^2.$$

In terms of the generating function $G(s,t)$, show that

$$\mathbf{V}(X(t)) = \left[\frac{\partial}{\partial s}\left(s\frac{\partial G(s,t)}{\partial s}\right) - \left(\frac{\partial G(s,t)}{\partial s}\right)^2 \right]_{s=1}$$

(an alternative formula to Eqn (5.20)). Obtain the variance for the Poisson process using its generating function

$$G(s,t) = e^{\lambda(s-1)t}$$

given by Eqn (5.17), and check your answer with that given in Problem 5.3.

5.5. A telephone answering service receives calls whose frequency varies with time but independently of other calls, perhaps with a daily pattern—more during the day than the night. The rate $\lambda(t) \geq 0$ becomes a function of the time t. The probability that a call arrives in the small time interval $(t, t + \delta t)$ when n calls have been received at time t satisfies

$$p_n(t + \delta t) = p_{n-1}(t)(\lambda(t)\delta t + o(\delta t)) + p_n(t)(1 - \lambda(t)\delta t + o(\delta t)), \qquad (n \geq 1),$$

with

$$p_0(t + \delta t) = (1 - \lambda(t)\delta t + o(\delta t))p_0(t).$$

It is assumed that the probability of two or more calls arriving in the interval $(t, t + \delta t)$ is negligible. Find the set of differential-difference equations for $p_n(t)$. Obtain the probability generating function $G(s,t)$ for the process and confirm that it is a stochastic process with parameter $\int_0^t \lambda(x)dx$. Find $p_n(t)$ by expanding $G(s,t)$ in powers of s. What is the mean number of calls received at time t?

5.6. For the telephone answering service in Problem 5.5, suppose that the rate is periodic given by $\lambda(t) = a + b\cos(\omega t)$ where $a > 0$ and $|b| < a$. Using the probability generating function from Problem 5.5, find the probability that n calls have been received at time t. Find also the mean number of calls received at time t. Sketch graphs of $p_0(t)$, $p_1(t)$, and $p_2(t)$ when $a = 0.5$, $b = 0.2$, and $\omega = 1$.

5.7. A Geiger counter is pre-set so that its initial reading is n_0 at time $t = 0$. What are the initial conditions on $p_n(t)$, the probability that the reading is $n(\geq n_0)$ at time t, and its generating function $G(s,t)$? Find $p_n(t)$, and the mean reading of the counter at time t.

5.8. A Poisson process with random variable $N(t)$ has probabilities

$$p_n(t) = \mathbf{P}[N(t) = n] = \frac{(\lambda t)^n e^{-\lambda t}}{n!}.$$

If $\lambda = 0.5$, calculate the following probabilities associated with the process:

(a) $\mathbf{P}[N(3) = 6]$;

(b) $\mathbf{P}[N(2.6) = 3]$;

(c) $\mathbf{P}[N(3.7) = 4|N(2.1) = 2]$;

(d) $\mathbf{P}[N(7) - N(3) = 3]$.

5.9. A telephone banking service receives an average of 1,000 calls per hour. On average a customer transaction takes one minute. If the calls arrive as a Poisson process, how many operators should the bank employ to avoid an expected accumulation of incoming calls?

5.10. A Geiger counter automatically switches off when the nth particle has been recorded, where n is fixed. The arrival of recorded particles is assumed to be a Poisson process with parameter λ. What is the expected value of the switch-off times?

5.11. Particles are emitted from a radioactive source, and $N(t)$, the random variable of the number of particles emitted up to time t from $t = 0$, is a Poisson process with intensity λ. The probability that any particle hits a certain target is p, independently of any other particle. If $M(t)$ is the random variable of the number of particles that hit the target up to time t, show, using the law of total probability, that $M(t)$ forms a Poisson process with intensity λp.

CHAPTER 6

Birth and Death Processes

6.1 Introduction

We shall now continue our investigation of further random processes in continuous time. In the Poisson process in Chapter 5, the probability of a further event was independent of the current number of events or readings in the Geiger counter analogy: this is a specific assumption in the definition of the Poisson process (Section 5.7). On the other hand, in birth and death processes, the probability of a birth or death will depend on the population size at time t. The more individuals in the population, the greater the possibility of a death, for example. As for Markov chains, birth and death processes are further examples of **Markov processes**. Additionally they include queueing processes, epidemics, predator–prey competition, and others. Markov processes are characterized by the condition that future development of these processes depends only on their current states and not their history up to that time. Generally Markov processes are easier to model and analyse, and they do include many interesting applications. Non-Markov processes in which the future state of a process depends on its whole history are generally harder to analyse mathematically.

We have adopted a gradual approach to the full problem. The birth and death processes are looked at separately, and then combined into the full birth and death process. Generally the partition theorem approach is used to derive the equations for $p_n(t)$, the probability that the population size is n at time t, and the probability generating function approach is the preferred method of solution.

6.2 The birth process

In this process everyone lives forever: there are no deaths. This process could model a colony of bacteria in which each cell randomly and independently divides into two cells at some future time, and the same happens for each divided cell. The births could start with n_0 cells at time $t = 0$. We shall assume in this first model that the probability that any individual cell divides in the time interval $(t, t + \delta t)$ is proportional to the time interval δt for small δt. If λ is the **birth rate** associated with this process, then the probability that the cell divides is $\lambda \delta t$ in the interval. For n cells the probability of cell division is $\lambda n \delta t$. To avoid complications, we assume that the probability that two or more births take place in the time interval δt is $o(\delta t)$ (that is, it can be ignored: see Section 5.3) with the consequence that the probability that no cell divides is $1 - \lambda n \delta t - o(\delta t)$.

There are many possibilities which are excluded in this model, including multiple births (twins, triplets, etc.) which may be significant in any real situation. The probability that a cell divides may not be homogeneous in time—it could decline with time, for example; the probability could also depend both on the state of the host and the number of cells, in which case the parameter λ could be a function of the population size n. However, the *simple* birth process described above is an interesting starting point for studying a stochastic model of growth. It is also known as the **Yule[1] process**, named after one of its originators.

This birth process is an example of a continuous-time Markov chain (see Chapter 4) with an unbounded set of states $E_{n_0}, E_{n_0+1}, E_{n_0+2}, \ldots$, where E_n is the state in which the population size is n. In this chain, however, transitions can only take place between n and $n+1$ in time δt since there are no deaths. The probability that a transition occurs from E_n to E_{n+1} in time δt is approximately $\lambda n \delta t$ and that no transition occurs is approximately $1 - \lambda n \delta t$. In a continuous-time Markov chain, the chain may spend varying periods of time in any state as the population grows.

If $N(t)$ is the random variable associated with the process, then we write

$$\mathbf{P}\{N(t) = n\} = p_n(t),$$

where $p_n(t)$ is the probability that the population size is n at time t. If the initial population size is $n_0 \geq 1$ at time $t = 0$, then

$$p_{n_0}(0) = 1, \text{ and } p_n(0) = 0 \text{ for } n > n_0. \tag{6.1}$$

According to the rules outlined above, a population of size n at time $t + \delta t$ can arise either from a population of size $n - 1$ at time t with a birth occurring with probability $\lambda(n-1)\delta t + o(\delta t)$ or through no event when the population is n, which can occur with probability $1 - \lambda n \delta t + o(\delta t)$, so that for $n \geq n_0 + 1$,

$$p_n(t + \delta t) = p_{n-1}(t)[\lambda(n-1)\delta t + o(\delta t)] + p_n(t)[1 - \lambda n \delta t + o(\delta t)], \quad (n \geq n_0 + 1)$$

(for explanations of the notations $o(\delta t)$ and $o(1)$, see the Appendix). This equation can be re-arranged into

$$\frac{p_n(t + \delta t) - p_n(t)}{\delta t} = \lambda(n-1)p_{n-1}(t) - \lambda n p_n(t) + o(1).$$

For the special case $n = n_0$,

$$p_{n_0}(t + \delta t) = p_{n_0}(t)[1 - \lambda n_0 \delta t + o(\delta t)],$$

which is equivalent to

$$\frac{p_{n_0}(t + \delta t) - p_{n_0}(t)}{\delta t} = -\lambda n_0 p_{n_0}(t) + o(1).$$

As $\delta t \to 0$ in both ratios, they become derivatives in the limit so that

$$\frac{dp_{n_0}(t)}{dt} = -\lambda n_0 p_{n_0}(t), \tag{6.2a}$$

[1] George Udny Yule (1871–1951), Scottish statistician.

$$\frac{dp_n(t)}{dt} = \lambda(n-1)p_{n-1}(t) - \lambda n p_n(t), \quad (n \geq n_0 + 1). \tag{6.2b}$$

These are differential-difference equations for the simple birth process. Since this is a *birth* process it follows that $p_n(t) = 0$ for $n < n_0$.

The system of Eqn (6.2) can be solved successively starting with $n = n_0$. The first equation

$$\frac{dp_{n_0}(t)}{dt} = -\lambda n_0 p_{n_0}(t)$$

has the solution

$$p_{n_0}(t) = e^{-\lambda n_0 t},$$

since $p_{n_0}(0) = 1$. Put $n = n_0 + 1$ in Eqn (6.2). Then the next equation is

$$\frac{dp_{n_0+1}(t)}{dt} + \lambda(n_0 + 1)p_{n_0+1}(t) = \lambda n_0 p_{n_0}(t) = \lambda n_0 e^{-\lambda n_0 t},$$

which is a standard first-order equation of the integrating-factor type. This can be solved to give

$$p_{n_0+1}(t) = n_0 e^{-\lambda n_0 t}(1 - e^{-\lambda t}),$$

since $p_{n_0+1}(0) = 0$. This process can be continued but it becomes tedious. It is easier to use the probability generating function method, which is also more general in its applicability. The generating function was first introduced in Section 1.9, and used also in Section 5.5 for the Poisson process.

Consider the probability generating function

$$G(s,t) = \sum_{n=n_0}^{\infty} p_n(t)s^n.$$

Multiply both sides of (6.2a) by s^{n_0}, and both sides of (6.2b) by s^n for $n \geq n_0 + 1$. Sum the equations for $n \geq n_0$. The result is

$$\sum_{n=n_0}^{\infty} \frac{dp_n(t)}{dt}s^n = \lambda \sum_{n=n_0+1}^{\infty}(n-1)p_{n-1}(t)s^n - \lambda \sum_{n=n_0}^{\infty} n p_n(t)s^n. \tag{6.3}$$

Consider each of the series in (6.3):

$$\sum_{n=n_0}^{\infty} \frac{dp_n(t)}{dt}s^n = \frac{\partial G(s,t)}{\partial t}, \tag{6.4}$$

$$\lambda \sum_{n=n_0+1}^{\infty}(n-1)p_{n-1}(t)s^n = \lambda \sum_{m=n_0}^{\infty} m p_m(t)s^{m+1} = \lambda s^2 \frac{\partial G(s,t)}{\partial s}, \tag{6.5}$$

$$\lambda \sum_{n=n_0}^{\infty} n p_n(t)s^n = \lambda s \frac{\partial G(s,t)}{\partial s}. \tag{6.6}$$

Equation (6.3) can be replaced therefore by the *partial differential equation*

$$\frac{\partial G(s,t)}{\partial t} = \lambda s^2 \frac{\partial G(s,t)}{\partial s} - \lambda s \frac{\partial G(s,t)}{\partial s} = \lambda s(s-1)\frac{\partial G(s,t)}{\partial s}. \tag{6.7}$$

We now have to solve this generating function equation. The initial condition $p_{n_0}(0) = 1$ translates into

$$G(s, 0) = s^{n_0} \tag{6.8}$$

for the generating function.

6.3 Birth process: Generating function equation

We shall look at the solution of Eqns (6.7) and (6.8) in some detail since the method of solution is used in other applications. A change of variable is applied to Eqn (6.7) to remove the term $\lambda s(s - 1)$. We achieve this by putting

$$\frac{ds}{dz} = \lambda s(s - 1), \tag{6.9}$$

and we shall assume that $0 < s < 1$ (for convergence reasons we are interested in the series for $G(s, t)$ for small s). This is a *first-order separable equation* which can be separated and integrated as follows:

$$\int \frac{ds}{s(1 - s)} = -\int \lambda dz = -\lambda z$$

(the value of any constant of integration is immaterial so we set it to a convenient value, in this case, to zero). Partial fractions are required on the left-hand side. Thus

$$\int \left[\frac{1}{s} + \frac{1}{1 - s} \right] ds = -\lambda z, \quad \text{or} \quad \ln \left[\frac{s}{1 - s} \right] = -\lambda z.$$

The solution of this equation for s gives

$$\frac{s}{1 - s} = e^{-\lambda z} \quad \text{or} \quad s = \frac{1}{1 + e^{\lambda z}}. \tag{6.10}$$

Let

$$Q(z, t) = G(s, t) = G(1/(1 + e^{\lambda z}), t).$$

We can check that

$$\frac{\partial Q(z, t)}{\partial z} = \frac{\partial G(1/(1 + e^{\lambda z}), t)}{\partial z} = \frac{\partial G(s, t)}{\partial s} \cdot \frac{ds}{dz} = \lambda s(s - 1) \frac{\partial G(s, t)}{\partial s},$$

using the chain rule in differentiation. Thus Eqn (6.7) becomes the simpler partial differetial equation

$$\frac{\partial Q(z, t)}{\partial t} = \frac{\partial Q(z, t)}{\partial z}.$$

The general solution of this equation for $Q(z, t)$ is any differentiable function, w, say, of $z + t$, that is, $Q(z, t) = w(z + t)$. This can be verified since

$$\frac{\partial}{\partial z} Q(z, t) = \frac{\partial}{\partial z} w(z + t) = w'(z + t) \cdot \frac{\partial(z + t)}{\partial z} = w'(z + t),$$

$$\frac{\partial}{\partial t} Q(z, t) = \frac{\partial}{\partial t} w(z + t) = w'(z + t) \cdot \frac{\partial(z + t)}{\partial t} = w'(z + t),$$

where $w'(z+t)$ means $dw(z+t)/d(z+t)$, in other words the derivative of w with respect to its argument. Note that if w is a function of $z+t$, then

$$w'(z+t) = \frac{\partial w(z+t)}{\partial z} = \frac{\partial w(z+t)}{\partial t}.$$

The function w is determined by the initial conditions. Thus, from Eqns (6.8) and (6.10),

$$G(s,0) = s^{n_0} = \frac{1}{(1+e^{\lambda z})^{n_0}} = w(z) = Q(z,0). \tag{6.11}$$

The function is defined in the middle of Eqn (6.11). Thus by changing the argument in w from z to $z+t$, and then back to s, we obtain

$$\begin{aligned}
G(s,t) &= Q(z,t) = w(z+t) = \frac{1}{[1+e^{\lambda(z+t)}]^{n_0}} \\
&= \frac{1}{[1+\frac{(1-s)}{s}e^{\lambda t}]^{n_0}} = \frac{s^{n_0}e^{-\lambda n_0 t}}{[1-(1-e^{-\lambda t})s]^{n_0}}, \tag{6.12}
\end{aligned}$$

using the change of variable Eqn (6.10) again. The probability generating function for the simple birth process is given by Eqn (6.12). The individual probabilities are the coefficients of s^n in the power series expansion of Eqn (6.12), which can be obtained by applying the binomial theorem to the denominator to derive a power series in s. Thus

$$\begin{aligned}
G(s,t) &= \frac{s^{n_0}e^{-\lambda n_0 t}}{[1-(1-e^{-\lambda t})s]^{n_0}} \\
&= s^{n_0}e^{-\lambda n_0 t}\left[1+\frac{n_0}{1!}(1-e^{-\lambda t})s + \frac{n_0(n_0+1)}{2!}(1-e^{-\lambda t})^2 s^2 + \cdots\right] \\
&= s^{n_0}e^{-\lambda n_0 t}\sum_{m=0}^{\infty}\binom{m+n_0-1}{n_0-1}(1-e^{-\lambda t})^m s^m, \\
&= e^{-\lambda n_0 t}\sum_{n=n_0}^{\infty}\binom{n-1}{n_0-1}(1-e^{-\lambda t})^{n-n_0}s^n, \tag{6.13}
\end{aligned}$$

putting $m = n - n_0$, where

$$\binom{r}{s} = \frac{r!}{s!(r-s)!}$$

is the binomial coefficient. The binomial coefficient with $s = 0$ is

$$\binom{r}{0} = \frac{r!}{0!r!} = \frac{1}{0!} = 1,$$

for all $r \geq 0$, since $0!$ is defined to be 1. Hence if $n_0 = 1$, then

$$\binom{n-1}{0} = 1,$$

for all $n \geq 1$. Finally from (6.13) the coefficients of the powers of s imply that, since the leading power is s^{n_0},

$$p_n(t) = 0, \qquad (n < n_0),$$

$$p_n(t) = \binom{n-1}{n_0-1} e^{-\lambda n_0 t}(1 - e^{-\lambda t})^{n-n_0}, \qquad (n \geq n_0),$$

which is a Pascal distribution with parameters $(n_0, e^{-\lambda t})$ (see Section 1.7). Some graphs of the first four probabilities are shown in Figure 6.1 for the birth process, starting with just one individual, $n_0 = 1$.

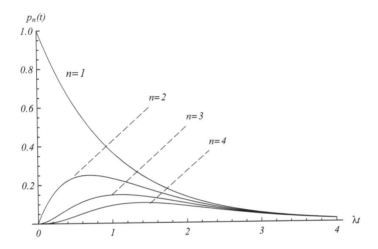

Figure 6.1 *The probabilities $p_n(t)$ shown for an initial population $n_0 = 1$ and $n = 1, 2, 3, 4$: a dimensionless time scale $\tau = \lambda t$ has been used.*

The mean population size at time t is given by (Eqns (5.18) and (6.12))

$$
\begin{aligned}
\mu(t) &= G_s(1,t) = \frac{\partial}{\partial s}\left[\frac{s^{n_0} e^{-\lambda n_0 t}}{[1 - (1 - e^{-\lambda t})s]^{n_0}} \right]_{s=1} \\
&= \left[\frac{n_0 s^{n_0-1} e^{-\lambda n_0 t}}{[1 - (1 - e^{-\lambda t})s]^{n_0}} + \frac{n_0 s^{n_0} e^{-\lambda n_0 t}(1 - e^{-\lambda t})}{[1 - (1 - e^{-\lambda t})s]^{n_0+1}} \right]_{s=1} \\
&= n_0(1 - e^{-\lambda t})e^{\lambda t} + n_0 = n_0 e^{\lambda t},
\end{aligned}
$$

or it can be deduced directly from the mean of the Pascal distribution. The expected population size grows exponentially with time.

6.4 The death process

In this case there are no births and the population numbers decline through deaths. Again we assume that the probability that any individual dies in a short time interval δt is $\mu \delta t$ where μ is the death rate. The probability that a death occurs in a population of size n is $\mu n \delta t$ and as before the probability of multiple deaths in interval δt is assumed to be negligible. By arguments similar to those for the birth process,

$$p_0(t + \delta t) = [\mu \delta t + o(\delta t)]p_1(t) + [1 + o(\delta t)]p_0(t),$$

$$p_n(t+\delta t) = [\mu(n+1)\delta t + o(\delta t)]p_{n+1}(t) + [1 - \mu n\delta t - o(\delta t)]p_n(t), \quad (1 \le n \le n_0 - 1).$$

If the initial population size is n_0, then $p_n(t) = 0$ for $n > n_0$ for all t, and since this is a death process,

$$p_{n_0}(t + \delta t) = [1 - \mu n_0 \delta t + o(\delta t)]p_{n_0}(t).$$

Thus, after rearrangement, the three previous equations can be expressed in the forms

$$\frac{p_0(t + \delta t) - p_0(t)}{\delta t} = \mu p_1(t) + o(1),$$

$$\frac{p_n(t + \delta t) - p_n(t)}{\delta t} = \mu(n + 1)p_{n+1}(t) - \mu n p_n(t) + o(1), \qquad (1 \le n \le n_0 - 1),$$

$$\frac{p_{n_0}(t + \delta t) - p_{n_0}(t)}{\delta t} = -\mu n_0 p_{n_0}(t) + o(1).$$

Now let $\delta t \to 0$ to obtain the differential-difference equations for the death process:

$$\left. \begin{array}{c} \dfrac{dp_0(t)}{dt} = \mu p_1(t), \\[2mm] \dfrac{dp_n(t)}{dt} = \mu(n + 1)p_{n+1}(t) - \mu n p_n(t) \quad (1 \le n \le n_0 - 1) \\[2mm] \dfrac{dp_{n_0}(t)}{dt} = -\mu n_0 p_{n_0}(t) \end{array} \right\} . \qquad (6.14)$$

If the initial population size is n_0 at time $t = 0$, then $p_{n_0}(0) = 1$. We expect the probability generating function to be a *finite series*. Multiply each equation in Eqn (6.14) by s^n as appropriate, and sum over $0 \le n \le n_0$:

$$\sum_{n=0}^{n_0} \frac{dp_n(t)}{dt} s^n = \mu \sum_{n=0}^{n_0-1} (n + 1)p_{n+1}(t)s^n - \mu \sum_{n=1}^{n_0} n p_n(t)s^n. \qquad (6.15)$$

Define the probability generating function $G(s, t)$ as

$$G(s, t) = \sum_{n=0}^{n_0} p_n(t)s^n.$$

Then the left-hand side of Eqn (6.15) is $\partial G(s, t)/\partial t$, and the two series on the right-hand side can be represented by

$$\sum_{n=0}^{n_0-1} (n + 1)p_{n+1}(t)s^n = \frac{\partial G(s, t)}{\partial s} \qquad \sum_{n=0}^{n_0} n p_n(t)s^n = s\frac{\partial G(s, t)}{\partial s}.$$

Finally $G(s, t)$ satisfies the partial differential equation

$$\frac{\partial G(s, t)}{\partial t} = \mu \frac{\partial G(s, t)}{\partial s} - \mu s \frac{\partial G(s, t)}{\partial s} = \mu(1 - s)\frac{\partial G(s, t)}{\partial s}. \qquad (6.16)$$

The method of solution is similar to that given for the birth process in the previous section. The difference between Eqns (6.7) and (6.16) lies solely in a different change of variable:

$$\frac{ds}{dz} = \mu(1 - s).$$

On this occasion,

$$\int \frac{ds}{1 - s} = \int \mu dz = \mu z.$$

Hence, for $0 < s < 1$,

$$- \ln(1 - s) = \mu z, \qquad \text{or} \qquad s = 1 - e^{-\mu z}.$$

For the death process we let

$$Q(z, t) = G(s, t) = G(1 - e^{-\mu z}, t),$$

where $Q(z, t)$ now satisfies

$$\frac{\partial Q(z, t)}{\partial t} = \frac{\partial Q(z, t)}{\partial z}.$$

which is the same partial differential equation as for the birth process but with a *different* change of variable. As before, the general solution is

$$G(s, t) = w(t + z).$$

If the initial population size is n_0, then $G(s, 0) = s^{n_0}$. Note how the initial value n_0 appears in the generating function. Then

$$G(s, 0) = s^{n_0} = (1 - e^{-\mu z})^{n_0} = w(z) = Q(z, 0).$$

Hence,

$$
\begin{aligned}
G(s, t) &= Q(z, t) = w(z + t) = (1 - e^{-\mu(z+t)})^{n_0} \\
&= [1 - e^{-\mu t}(1 - s)]^{n_0} \\
&= (1 - e^{-\mu t})^{n_0} \left[1 + \frac{se^{-\mu t}}{1 - e^{-\mu t}} \right]^{n_0} \\
&= \sum_{n=0}^{n_0} \binom{n_0}{n} e^{-n\mu t} (1 - e^{-\mu t})^{n_0 - n} s^n
\end{aligned}
\tag{6.17}
$$

where the series is obtained using the binomial theorem. The individual probabilities are

$$p_n(t) = \binom{n_0}{n} e^{-n\mu t} (1 - e^{-\mu t})^{n_0 - n},$$

for $n = 0, 1, 2, \ldots, n_0$[2], which are binomially distributed.

Example 6.1. *Find the mean size $\mu(t)$ of the population in the death process at time t. Show that $\mu(t)$ satisfies the differential equation*

$$\frac{d\mu(t)}{dt} = -\mu\mu(t),$$

where $p_{n_0}(0) = 1$. Interpret the result.

---

[2] Remember, the binomial coefficent $\binom{n_0}{0} = 1$.

The mean is given by

$$
\begin{aligned}
\mu(t) &= G_s(1,t) = \frac{\partial}{\partial s}\left[1 - e^{-\mu t}(1-s)\right]^{n_0}\bigg|_{s=1} \\
&= n_0 e^{-\mu t}\left[1 - e^{-\mu t}(1-s)\right]^{n_0-1}\big|_{s=1} \\
&= n_0 e^{-\mu t}.
\end{aligned}
$$

Hence,

$$
\frac{d\mu(t)}{dt} = -\mu n_0 e^{-\mu t} = -\mu\mu(t),
$$

with $\mu(0) = n_0$ as required.

Suppose that we consider a **deterministic model**[3] of a population in which the population $n(t)$ is a continuous function of time rather than a random variable with *discrete* values. This is a justifiable approximation at least for large populations. We could then model the population change by postulating that the rate of decrease of the population is proportional to the current population size $n(t)$. Thus

$$
\frac{dn(t)}{dt} \propto n(t) \qquad \text{or} \qquad \frac{dn(t)}{dt} = -\mu n(t)
$$

where μ is the death-rate. This is the **Malthus**[4] **model** for the death process. We can deduce from this that the mean of the stochastic process satisfies the differential equation of the simple deterministic model. This provides some justification for using deterministic models in large populations.

It is easy to calculate from the probability generating function the probability of **extinction** at time t. It is the probability that the population size is zero at time t, namely,

$$
p_0(t) = G(0,t) = [1 - e^{-\mu t}]^{n_0}.
$$

The probability of **ultimate extinction** is

$$
\lim_{t\to\infty} p_0(t) = \lim_{t\to\infty} G(0,t) = \lim_{t\to\infty} [1 - e^{-\mu t}]^{n_0} = 1,
$$

since $\lim_{t\to\infty} e^{-\mu t} = 0$. In other words the probability of ultimate extinction is certain, as we would expect in a death process.

6.5 The combined birth and death process

Both processes of Sections 6.2 and 6.4 are now combined into one with a birth rate λ and a death rate μ, although this is not achieved by simple addition of the equations. However, using similar arguments as to how a population of size n can arise at time $t + \delta t$, we obtain

$$
p_0(t + \delta t) = [\mu\delta t + o(\delta t)]p_1(t) + [1 + o(\delta t)]p_0(t),
$$

$$
\begin{aligned}
p_n(t + \delta t) &= [\lambda(n-1)\delta t + o(\delta t)]p_{n-1}(t) + [1 - (\lambda n + \mu n)\delta t + o(\delta t)]p_n(t) \\
&\quad + [\mu(n+1)\delta t + o(\delta t)]p_{n+1}(t), \qquad (n \geq 1).
\end{aligned}
$$

---

[3] See Section 9.9 for more about deterministic models

[4] Thomas Malthus (1766–1834), English economist/statistician: well-known for his work on population growth.

In the limit $\delta t \to 0$, $p_n(t)$ satisfies

$$\left.\begin{array}{l} \dfrac{dp_0(t)}{dt} = \mu p_1(t) \\[2mm] \dfrac{dp_n(t)}{dt} = \lambda(n-1)p_{n-1}(t) - (\lambda+\mu)np_n(t) + \mu(n+1)p_{n+1}(t), \ (n \geq 1) \end{array}\right\}. \quad (6.18)$$

The value $\lambda = 0$ in (6.18) results in the death equations (6.14) whilst the choice $\mu = 0$ gives the birth equations (6.2a,b), although the range of n has to be carefully defined in both cases. Since births occur the probability generating function will be an *infinite* series defined by

$$G(s,t) = \sum_{n=0}^{\infty} p_n(t)s^n.$$

Multiplying the appropriate equations in Eqn (6.18) by s^n and summing over $n \geq 0$ we obtain (see Sections 6.2 and 6.4 for the sums of the series):

$$\begin{aligned} \frac{\partial G(s,t)}{\partial t} &= \lambda s(s-1)\frac{\partial G(s,t)}{\partial s} + \mu(1-s)\frac{\partial G(s,t)}{\partial s} \\[2mm] &= (\lambda s - \mu)(s-1)\frac{\partial G(s,t)}{\partial s}. \end{aligned} \quad (6.19)$$

There are two cases to consider, namely $\lambda \neq \mu$ and $\lambda = \mu$.

(a) $\lambda \neq \mu$

For the birth and death process, the required change of variable is given by the differential equation

$$\frac{ds}{dz} = (\lambda s - \mu)(s-1).$$

This simple separable equation can be integrated to give

$$\begin{aligned} z = \int dz &= \int \frac{ds}{(\lambda s - \mu)(s-1)} = \frac{1}{\lambda}\int \frac{ds}{(\frac{\mu}{\lambda}-s)(1-s)} \\[2mm] &= \frac{1}{\lambda-\mu}\int \left[\frac{1}{\frac{\mu}{\lambda}-s} - \frac{1}{1-s}\right]ds, \qquad \text{(using partial fractions)} \\[2mm] &= \frac{1}{\lambda-\mu}\ln\left[\frac{1-s}{\frac{\mu}{\lambda}-s}\right], \qquad (0 < s < \min(1, \tfrac{\mu}{\lambda})). \quad (6.20) \end{aligned}$$

The inversion of this formula defines s as

$$s = \frac{\lambda - \mu e^{(\lambda-\mu)z}}{\lambda - \lambda e^{(\lambda-\mu)z}}. \quad (6.21)$$

After the change of variable, let $Q(z,t) = G(s,t)$, so that $Q(z,t)$ satisfies

$$\frac{\partial Q(z,t)}{\partial t} = \frac{\partial Q(z,t)}{\partial z},$$

which has the general solution $Q(z,t) = w(z+t)$.

If the initial population size is n_0, then

$$G(s,0) = s^{n_0} = \left[\frac{\lambda - \mu e^{(\lambda-\mu)z}}{\lambda - \lambda e^{(\lambda-\mu)z}}\right]^{n_0} = w(z) = Q(z,0).$$

Hence,

$$G(s,t) = Q(z,t) = w(z+t) = \left[\frac{\lambda - \mu e^{(\lambda-\mu)(z+t)}}{\lambda - \lambda e^{(\lambda-\mu)(z+t)}}\right]^{n_0}. \tag{6.22}$$

From (6.20),

$$e^{(\lambda-\mu)z} = \frac{1-s}{\frac{\mu}{\lambda}-s} = \frac{\lambda(1-s)}{\mu-\lambda s}.$$

Finally, elimination of z in Eqn (6.22) leads to

$$G(s,t) = \left[\frac{\mu(1-s) - (\mu - \lambda s)e^{-(\lambda-\mu)t}}{\lambda(1-s) - (\mu - \lambda s)e^{-(\lambda-\mu)t}}\right]^{n_0}. \tag{6.23}$$

The expansion of $G(s,t)$ as a power series in s is the product of two binomial series, which can be found, but it is complicated.

The expected population size is, at time t, for $\lambda \neq \mu$,

$$\begin{aligned}
\mu(t) &= \sum_{n=1}^{\infty} np_n(t) = G_s(1,t) \\
&= \frac{n_0(-\mu + \lambda e^{-(\lambda-\mu)t})}{-(\mu-\lambda)e^{-(\lambda-\mu)t}} - \frac{n_0(-\lambda + \lambda e^{(\lambda-\mu)t})}{-(\mu-\lambda)e^{-(\lambda-\mu)t}} \\
&= n_0 e^{(\lambda-\mu)t}.
\end{aligned}$$

(b) $\lambda = \mu$, (birth and death rates the same).

In this case Eqn (6.19) becomes

$$\frac{\partial G(s,t)}{\partial t} = \lambda(1-s)^2 \frac{\partial G(s,t)}{\partial s}.$$

Let

$$\frac{ds}{dz} = \lambda(1-s)^2.$$

Then the change of variable is

$$z = \frac{1}{\lambda}\int \frac{ds}{(1-s)^2} = \frac{1}{\lambda(1-s)} \quad \text{or} \quad s = \frac{\lambda z - 1}{\lambda z}.$$

It follows that

$$w(z) = s^{n_0} = \left[\frac{\lambda z - 1}{\lambda z}\right]^{n_0}.$$

Finally, the probability generating function for the special case in which the birth and death rates are the same is given by

$$G(s,t) = \left[\frac{\lambda(z+t) - 1}{\lambda(z+t)}\right]^{n_0} = \left[\frac{1 + (\lambda t - 1)(1-s)}{1 + \lambda t(1-s)}\right]^{n_0}. \tag{6.24}$$

The expected value of the population size at time t in the case $\lambda = \mu$ is left as a problem at the end of the chapter. The probability of extinction at time t for the case $\lambda = \mu$ is, from (6.24),

$$p_0(t) = G(0, t) = \left[\frac{1 + (\lambda t - 1)}{1 + \lambda t}\right]^{n_0} = \left[\frac{\lambda t}{1 + \lambda t}\right]^{n_0}. \qquad (6.25)$$

As t becomes large this value approaches 1 since

$$\lim_{t \to \infty} p_0(t) = \lim_{t \to \infty} \left[\frac{1}{1 + \frac{1}{\lambda t}}\right]^{n_0} = 1.$$

We obtain the interesting conclusion that, if the birth and death rates are in balance, then ultimate extinction is certain.

That the probability generating functions in Eqns (6.23) and (6.24) are powers of certain functions of s and t is not surprising if we recollect the definition of the generating function (see Section 1.9) as

$$G(s, t) = \mathbf{E}[s^{N(t)}],$$

where $N(t)$ is a random variable of the number of individuals in the population at time t. Suppose that we identify the n_0 individuals of the initial population and let $N_i(t)$ represent the number of descendants of the i-th individual so that

$$N(t) = N_1(t) + N_2(t) + \cdots + N_{n_0}(t).$$

Hence,

$$G(s, t) = \mathbf{E}[s^{N(t)}] = \mathbf{E}[s^{N_1(t) + N_2(t) + \cdots + N_{n_0}(t)}].$$

Since the $N_i(t)$, $(i = 1, 2, \ldots, n_0)$ are independent and identically distributed (iid), it follows that

$$G(s, t) = \mathbf{E}[s^{N_1(t)}]\mathbf{E}[s^{N_2(t)}] \ldots \mathbf{E}[s^{N_{n_0}(t)}] = [\mathbf{E}[s^{N_1(t)}]]^{n_0},$$

where $N_1(t)$ has been chosen since the means are all the same values (see Section 1.9e). Hence $G(s, t)$ must be the n_0-th power of some function of s and t.

Example 6.2. *In the birth and death process with $\lambda = \mu$, show that the mean time to extinction is infinite for any initial population size n_0.*

We require the probability distribution function $F(t)$ for the time T_{n_0} to extinction, that is,

$$\begin{aligned}
F(t) &= \mathbf{P}\{T_{n_0} \le t\} \\
&= p_0(t) = \left[\frac{\lambda t}{1 + \lambda t}\right]^{n_0}
\end{aligned}$$

from (6.25). The density function $f(t)$ of T_{n_0} is

$$f(t) = \frac{dF(t)}{dt} = \frac{n_0 \lambda^{n_0} t^{n_0 - 1}}{(1 + \lambda t)^{n_0}} - \frac{n_0 (\lambda t)^{n_0} \lambda}{(1 + \lambda t)^{n_0 + 1}} = \frac{n_0 \lambda^{n_0} t^{n_0 - 1}}{(1 + \lambda t)^{n_0 + 1}}.$$

The expected value of the random variable T_{n_0} will be given by

$$\mathbf{E}(T_{n_0}) = \int_0^\infty t f(t) \, dt$$

only if the integral on the right *converges*. In this case

$$tf(t) = \frac{n_0 \lambda^{n_0} t^{n_0}}{(1 + \lambda t)^{n_0+1}}.$$

For *large* t, $1 + \lambda t \approx \lambda t$ and

$$tf(t) \approx \frac{n_0 \lambda^{n_0} t^{n_0}}{(\lambda t)^{n_0+1}} = \frac{n_0}{\lambda} \frac{1}{t}.$$

Although $tf(t) \to 0$ as $t \to \infty$, it is too slow for the integral to converge. For example, the integral

$$\int_1^\tau \frac{dt}{t} = \left[\ln t\right]_1^\tau = \ln \tau \to \infty$$

as $\tau \to \infty$ has the same behaviour for large t and this integral *diverges*. We conclude that $\mathbf{E}(T_{n_0}) = \infty$. Notwithstanding that extinction is ultimately certain, we expect it on average to take an infinite time. A graph of $p_0(t)$ is shown in Figure 6.2.

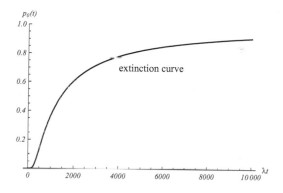

Figure 6.2 *The graph shows the behaviour of the $p_0(t) = [\lambda t/(1 + \lambda t)]^{n_0}$, the probability of extinction, against λt for an initial population size of $n_0 = 1,000$ in the case $\lambda = \mu$. The actual time scale will depend on the parameter λ.*

Finally in this section a computed stochastic output for a birth and death process will be shown. In any time interval δt when the population is n, the probability of a birth is $\lambda n \delta t$, of a death $\mu n \delta t$, and of neither is $1 - (\lambda - \mu) n \delta t$: only one of these events can occur. Figure 6.3 shows the discrete population sequence at times $m \delta t$ for $m = 0, 1, 2, \ldots, 200$ in this case. The birth and death rates $\lambda = 0.1$ and $\mu = 0.1$ are chosen together with the time step $\delta t = 0.05$. The initial population size is $n_0 = 50$. In the illustration in Figure 6.3 the mean is 63.9. It is possible for the population to take a long time to return to the initial population: at the end of this epoch the population is 90. If you have access to a program to create outputs for this symmetric model, it is instructive to run a sequence of trials with this or a larger number of time-steps to see how frequently a return to $n = 50$ occurs. This is essentially the same discussion as occurred for the symmetric random walk in Section 3.4.

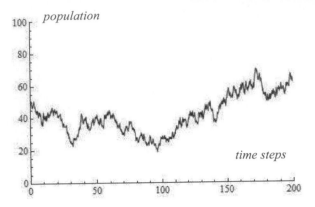

Figure 6.3 *Birth and death population with birth rate* $\lambda = 0.1$, *the same death rate* $\mu = 0.1$ *for 200 time steps of duration* $\delta t = 0.05$. *The initial population size is* $n_0 = 50$. *For this output the sample mean is* $\mu = 63.9$

6.6 General population processes

The previous models for birth and death processes assume that the rates of births and deaths are simply proportional to the population size. In more general population processes the rates can be more general functions of n, which perhaps could represent more realistic models such as higher death rates for overcrowded larger populations.

Assume that the birth and death rates are λ_n and μ_n, respectively. Using arguments which lead to Eqns (6.14) but with the new coefficients, the governing differential-difference equations are replaced by

$$\left.\begin{array}{l} \dfrac{dp_0(t)}{dt} = -\lambda_0 p_0(t) + \mu_1 p_1(t) \\[2mm] \dfrac{dp_n(t)}{dt} = \lambda_{n-1}p_{n-1}(t) - (\lambda_n + \mu_n)p_n(t) + \mu_{n+1}p_{n+1}(t), \ (n \geq 1) \end{array}\right\}, \quad (6.26)$$

but note that λ_0 must be included since we cannot assume it is zero. The simple birth and death process is given by the particular case in which $\lambda_n = n\lambda$ and $\mu_n = n\mu$ are *linear* functions of n. The Poisson process of the previous chapter corresponds to $\lambda_n = \lambda$ and $\mu_n = 0$.

The following example develops a population model in which the birth rate is constant, perhaps sustained by immigration, and the death rate is simple and given by $\mu_n = n\mu$. In other words there are no indigenous births.

Example 6.3. *In a population model, the immigration rate* $\lambda_n = \lambda$ *is a constant, and the death rate is* $\mu_n = n\mu$. *Assuming that the initial population size is* n_0, *find the probability generating function for this process, and find the mean population size at time t.*

Insert $\lambda_n = \lambda$ and $\mu_n = n\mu$ into (6.26), multiply by s^n, and sum over $n = 0, 1, 2, \ldots$. The result is

$$\sum_{n=0}^{\infty} \frac{dp_n(t)}{dt} s^n = \lambda \sum_{n=1}^{\infty} p_{n-1}(t)s^n - \lambda \sum_{n=0}^{\infty} p_n(t)s^n - \mu \sum_{n=1}^{\infty} np_n(t)s^n$$

$$+\mu \sum_{n=0}^{\infty} (n+1)p_{n+1}(t)s^n. \tag{6.27}$$

Using the general series for $G(s,t)$ (Section 6.5) and those given by (6.5) and (6.6) for its partial derivatives, Eqn (6.27) becomes

$$\begin{aligned}
\frac{\partial G(s,t)}{\partial t} &= \lambda s G(s,t) - \lambda G(s,t) - \mu s \frac{\partial G(s,t)}{\partial s} + \mu \frac{\partial G(s,t)}{\partial s} \\
&= \lambda(s-1)G(s,t) + \mu(1-s)\frac{\partial G(s,t)}{\partial s}.
\end{aligned} \tag{6.28}$$

This is a partial differential equation of a type we have not encountered previously since it includes the term $G(s,t)$ in addition to the partial derivatives. This term can be removed by introducing a new function $H(s,t)$ through the transformation

$$G(s,t) = e^{\lambda s/\mu}H(s,t).$$

Then

$$\frac{\partial G(s,t)}{\partial t} = e^{\lambda s/\mu}\frac{\partial H(s,t)}{\partial t}, \qquad \frac{\partial G(s,t)}{\partial s} = e^{\lambda s/\mu}\frac{\partial H(s,t)}{\partial s} + \frac{\lambda}{\mu}e^{\lambda s/\mu}H(s,t).$$

Substitution of the partial derivatives into (6.28) leads to

$$\frac{\partial H(s,t)}{\partial t} = \mu(1-s)\frac{\partial H(s,t)}{\partial s} \tag{6.29}$$

for $H(s,t)$. In fact $H(s,t)$ now satisfies the same equation as that for the death process in Section 6.4. However, the initial condition is different since, for an initial population of n_0,

$$G(s,0) = s^{n_0}, \quad \text{but} \quad H(s,0) = e^{-\lambda s/\mu}s^{n_0}.$$

The change of variable is

$$s = 1 - e^{-\mu z}.$$

Hence,

$$H(s,0) = e^{-\lambda s/\mu}s^{n_0} = e^{-\lambda(1-e^{-\mu z})/\mu}(1 - e^{-\mu z})^{n_0} = w(z), \text{ say,}$$

from which it follows that

$$\begin{aligned}
G(s,t) &= e^{\lambda s/\mu}w(z+t) \\
&= e^{\lambda s/\mu}\exp[-\lambda(1 - e^{-\mu(z+t)})/\mu][1 - e^{-\mu(z+t)}]^{n_0} \\
&= e^{\lambda s/\mu}\exp[-\lambda(1 - (1-s)e^{-\mu t})/\mu][1 - (1-s)e^{-\mu t}]^{n_0}. \tag{6.30}
\end{aligned}$$

To find the mean population size at time t, take the logarithm of both sides of this equation:

$$\ln G(s,t) = \frac{\lambda s}{\mu} - \frac{\lambda}{\mu}[1 - (1-s)e^{-\mu t}] + n_0 \ln[1 - (1-s)e^{-\mu t}].$$

Differentiate with respect to s and put $s = 1$:

$$\frac{G_s(1,t)}{G(1,t)} = \frac{\lambda}{\mu} - \frac{\lambda}{\mu}e^{-\mu t} + n_0 e^{-\mu t}.$$

Now $G(1,t) = \sum_{n=0}^{\infty} p_n(t) = 1$, and the mean is therefore

$$\mu(t) = G_s(1,t) = \frac{\lambda}{\mu}(1 - e^{-\mu t}) + n_0 e^{-\mu t}.$$

As $t \to \infty$, the long-term mean approaches the ratio of the rates λ/μ.

The birth and death rates can be functions of time in addition to population size. This might reflect declining fertility or seasonal variations in births or deaths. The setting up of the equation for the probability generating function presents no particular problem although its solution inevitably becomes more complicated. The following example illustrates a birth process with a declining birth rate.

Example 6.4. *A colony of bacteria grows without deaths with a birth rate which is time-dependent. The rate declines exponentially with time according to* $\lambda(t) = \alpha e^{-\beta t}$, *where* α *and* β *are constants. Find the mean population size at time* t *given that the initial size of the colony is* n_0.

For a time-varying birth rate the construction of the equation for the probability generating function is still given by (6.7) but with λ as a function of time. Variations of $\lambda(t)$ over an incremental time interval δt will have a lower effect on the probability of a birth. Thus $G(s,t)$ satisfies

$$\frac{\partial G(s,t)}{\partial t} = \lambda(t)s(s-1)\frac{\partial G(s,t)}{\partial s} = \alpha e^{-\beta t}s(s-1)\frac{\partial G(s,t)}{\partial s}$$

with $G(s,0) = s^{n_0}$. We now apply the *double* change of variable

$$\frac{ds}{dz} = \alpha s(s-1), \qquad \frac{dt}{d\tau} = e^{\beta t}.$$

As in Section 6.3,

$$s = \frac{1}{1 + e^{\alpha z}},$$

whilst

$$\tau = \frac{1}{\beta}(1 - e^{-\beta t})$$

so that $\tau = 0$ when $t = 0$. With these changes of variable,

$$Q(z,\tau) = G(s,t) = G(1/(1 + e^{\alpha z}), -\beta^{-1}\ln(1 - \beta\tau)),$$

where $Q(z,\tau)$ satisfies

$$\frac{\partial Q(z,\tau)}{\partial \tau} = \frac{\partial Q(z,\tau)}{\partial z}.$$

As in Section 6.3,

$$Q(z,\tau) = w(z + \tau)$$

for any differentiable function w. The initial condition determines w through

$$G(s,0) = s^{n_0} = \frac{1}{(1 + e^{\alpha z})^{n_0}} = w(z) = Q(z,0).$$

Hence,

$$
\begin{aligned}
G(s,t) &= Q(z,\tau) = w(z+\tau) = [1 + e^{\alpha(z+\tau)}]^{-n_0} \\
&= [1 + \left(\frac{1-s}{s}\right)e^{\alpha(1-e^{-\beta t})/\beta}]^{-n_0} \\
&= s^{n_0}[s + (1-s)e^{\alpha(1-e^{-\beta t})/\beta}]^{-n_0}.
\end{aligned}
\tag{6.31}
$$

The mean population size at time t is

$$\mu(t) = G_s(1,t) = \frac{\partial}{\partial s}\left[\frac{s^{n_0}}{[s + (1-s)e^{\alpha(1-e^{-\beta t})/\beta}]^{n_0}}\right]_{s=1} = n_0 e^{\alpha(1-e^{-\beta t})/\beta}.
\tag{6.32}$$

As $t \to \infty$, then the mean population approaches the limit $n_0 e^{\alpha/\beta}$ which, at first sight, seems surprising since there are no deaths. However, the reason for this limit is that the birth rate decreases exponentially to zero with time.

6.7 Problems

6.1. A colony of cells grows from a single cell. The probability that a cell divides in a time interval δt is

$$\lambda \delta t + o(\delta t).$$

There are no deaths. Show that the probability generating function for this birth process is

$$G(s,t) = \frac{se^{-\lambda t}}{1 - (1 - e^{-\lambda t})s}.$$

Find the probability that the original cell has not divided at time t, and the mean and variance of population size at time t (see Problem 5.4, for the variance formula using the probability generating function).

6.2. A simple birth process has a constant birth rate λ. Show that its mean population size $\mu(t)$ satisfies the differential equation

$$\frac{d\mu(t)}{dt} - \lambda \mu(t).$$

How can this result be interpreted in terms of a deterministic model for a birth process?

6.3. The probability generating function for a simple death process with death rate μ and initial population size n_0 is given by

$$G(s,t) = (1 - e^{-\mu t})^{n_0} \left[1 + \frac{se^{-\mu t}}{1 - e^{-\mu t}} \right]^{n_0}$$

(see Eqn (6.17)). Using the binomial theorem find the probability $p_n(t)$ for $n \leq n_0$. If n_0 is an *even* number, find the probability that the population size has halved by time t. A large number of experiments were undertaken with live samples with a variety of initial population sizes drawn from a common source and the times of the halving of deaths were recorded for each sample. What would be the expected time for the population size to halve?

6.4. A birth process has a probability generating function $G(s,t)$ given by

$$G(s,t) = \frac{s}{e^{\lambda t} + s(1 - e^{\lambda t})}.$$

(a) What is the initial population size?
(b) Find the probability that the population size is n at time t.
(c) Find the mean and variance of the population size at time t.

6.5. A random process has the probability generating function

$$G(s,t) = \left(\frac{2 + st}{2 + t} \right)^r,$$

where r is a positive integer. What is the initial state of the process? Find the probability $p_n(t)$

associated with the generating function. What is $p_r(t)$? Show that the mean associated with $G(s,t)$ is

$$\mu(t) = \frac{rt}{2+t}.$$

6.6. In a simple birth and death process with unequal birth and death rates λ and μ, the probability generating function is given by

$$G(s,t) = \left[\frac{\mu(1-s) - (\mu - \lambda s)e^{-(\lambda-\mu)t}}{\lambda(1-s) - (\mu - \lambda s)e^{-(\lambda-\mu)t}}\right]^{n_0},$$

for an initial population size n_0 (see Eqn (6.23)).
(a) Find the mean population size at time t.
(b) Find the probability of extinction at time t.
(c) Show that, if $\lambda < \mu$, then the probability of ultimate extinction is 1. What is the probability if $\lambda > \mu$?
(d) Find the variance of the population size.

6.7. In a population model, the immigration rate $\lambda_n = \lambda$, a constant, and the death rate $\mu_n = n\mu$. For an initial population size n_0, the probability generating function is (Example 6.3)

$$G(s,t) = e^{\lambda s/\mu} \exp[-\lambda(1 - (1-s)e^{-\mu t})/\mu][1 - (1-s)e^{-\mu t}]^{n_0}.$$

Find the probability that extinction occurs at time t.

6.8. In a general birth and death process a population is maintained by immigration at a constant rate λ, and the death rate is $n\mu$. Using the differential-difference equations (6.25) directly, obtain the differential equation

$$\frac{d\mu(t)}{dt} + \mu\mu(t) = \lambda,$$

for the mean population size $\mu(t)$. Solve this equation assuming an initial population n_0 and compare the answer with that given in Example 6.3.

6.9. In a death process the probability of a death when the population size is $n \neq 0$ is a constant μ but obviously zero if the population size is zero. Verify that, if the initial population is n_0, then $p_n(t)$, the probability that the population size is n at time t is given by

$$p_n(t) = \frac{(\mu t)^{n_0-n}}{(n_0-n)!}e^{-\mu t}, \qquad (1 \le n \le n_0),$$

$$p_0(t) = \frac{\mu^{n_0}}{(n_0-1)!}\int_0^t s^{n_0-1}e^{-\mu s}ds.$$

Show that the mean time to extinction is n_0/μ.

6.10. In a birth and death process the birth and death rates are given by

$$\lambda_n = n\lambda + \alpha, \qquad \mu_n = n\mu,$$

where α represents a constant immigration rate. Show that the probability generating function $G(s,t)$ of the process satisfies

$$\frac{\partial G(s,t)}{\partial t} = (\lambda s - \mu)(s-1)\frac{\partial G(s,t)}{\partial s} + \alpha(s-1)G(s,t).$$

Show also that, if

$$G(s,t) = (\mu - \lambda s)^{-\alpha/\lambda} H(s,t),$$

then $H(s,t)$ satisfies

$$\frac{\partial H(s,t)}{\partial t} = (\lambda s - \mu)(s - 1)\frac{\partial H(s,t)}{\partial s}.$$

Let the initial population size be n_0. Solve the partial differential equation for $H(s,t)$ using the method of Section 6.5 and confirm that

$$G(s,t) = \frac{(\mu - \lambda)^{\alpha/\lambda}[(\mu - \lambda s) - \mu(1 - s)e^{(\lambda-\mu)t}]^{n_0}}{[(\mu - \lambda s) - \lambda(1 - s)e^{(\lambda-\mu)t}]^{n_0+(\alpha/\lambda)}}.$$

(Remember the modified initial condition for $H(s,t)$.)

Find $p_0(t)$, the probability that the population is zero at time t (since immigration takes place even when the population is zero there is no question of extinction in this process). Hence show that

$$\lim_{t\to\infty} p_0(t) = \left(\frac{\mu - \lambda}{\mu}\right)^{\alpha/\lambda}$$

if $\lambda < \mu$. What is the limit if $\lambda > \mu$?

The long-term behaviour of the process for $\lambda < \mu$ can be investigated by looking at the limit of the probability generating function as $t \to \infty$. Show that

$$\lim_{t\to\infty} G(s,t) = \left(\frac{\mu - \lambda}{\mu - \lambda s}\right)^{\alpha/\lambda}.$$

This is the probability generating function of a **stationary distribution** and it indicates that a balance has been achieved between the birth and immigration rates, and the death rate. What is the long-term mean population size?

Obtain the probability generating function in the special case $\lambda = \mu$.

6.11. In a birth and death process with immigration, the birth and death rates are respectively

$$\lambda_n = n\lambda + \alpha, \qquad \mu_n = n\mu.$$

Show directly from the differential-difference equations for $p_n(t)$ that the mean population size $\mu(t)$ satisfies the differential equation

$$\frac{d\mu(t)}{dt} = (\lambda - \mu)\mu(t) + \alpha.$$

Deduce the result

$$\mu(t) \to \frac{\alpha}{\mu - \lambda}$$

as $t \to \infty$ if $\lambda < \mu$. Discuss the design of a deterministic immigration model based on this equation.

6.12. In a simple birth and death process with equal birth and death rates λ, the initial population size has a Poisson distribution with probabilities

$$p_n(0) = e^{-\alpha}\frac{\alpha^n}{n!}, \qquad (n = 0, 1, 2, \ldots)$$

with parameter α. It could be thought of as a process in which the initial distribution has arisen as the result of some previous process. Find the probability generating function for this process, and confirm that the probability of extinction at time t is $\exp[-\alpha/(1 + \lambda t)]$ and that the mean population size is α for all t.

6.13. A birth and death process takes place as follows. A single bacterium is allowed to grow and assumed to behave as a simple birth process with birth rate λ for a time t_1 without any deaths. No further growth then takes place. The colony of bacteria is then allowed to die with the assumption that it is a simple death process with death rate μ for a time t_2. Show that the probability of extinction after the total time $t_1 + t_2$ is

$$\sum_{n=1}^{\infty} e^{\lambda t_1}(1 - e^{-\lambda t_1})^{n-1}(1 - e^{-\mu t_2})^n.$$

Using the formula for the sum of a geometric series, show that this probability can be simplified to

$$\frac{e^{\mu t_2} - 1}{e^{\lambda t_1} + e^{\mu t_2} - 1}.$$

6.14. As in the previous problem a single bacterium grows as a simple birth process with rate λ and no deaths for a time τ. The colony numbers then decline as a simple death process with rate μ. Show that the probability generating function for the death process is

$$\frac{(1 - e^{-\mu t}(1 - s))e^{-\lambda \tau}}{1 - (1 - e^{-\lambda \tau})(1 - e^{-\mu t}(1 - s))},$$

where t is measured from the time τ. Show that the mean population size during the death process is $e^{\lambda \tau - \mu t}$.

6.15. For a simple birth and death process the probability generating function (Eqn (6.23)) is given by

$$G(s,t) = \left[\frac{\mu(1 - s) - (\mu - \lambda s)e^{-(\lambda - \mu)t}}{\lambda(1 - s) - (\mu - \lambda s)e^{-(\lambda - \mu)t}}\right]^{n_0}$$

for an initial population of n_0. What is the probability that the population size is (a) zero, (b) 1 at time t?

6.16. (An alternative method of solution for the probability generating function.) The general solution of the first-order partial differential equation

$$A(x,y,z)\frac{\partial z}{\partial x} + B(x,y,z)\frac{\partial z}{\partial y} = C(x,y,z)$$

is $f(u, v) = 0$, where f is an arbitrary function, and $u(x, y, z) = c_1$ and $v(x, y, z) = c_2$ are two independent solutions of

$$\frac{dx}{A(x,y,z)} = \frac{dy}{B(x,y,z)} = \frac{dz}{C(x,y,z)}.$$

This is known as **Cauchy's method**.

Apply the method to the partial differential equation for the probability generating function for the simple birth and death process, namely (Eqn (6.19))

$$\frac{\partial G(s,t)}{\partial t} = (\lambda s - \mu)(s - 1)\frac{\partial G(s,t)}{\partial s},$$

by solving

$$\frac{ds}{(\lambda s - \mu)(1 - s)} = \frac{dt}{1} = \frac{dG}{0}.$$

Show that

$$u(s, t, G) = G = c_1 \quad \text{and} \quad v(s, t, G) = e^{(\lambda - \mu)t} \left(\frac{1 - s}{\frac{\mu}{\lambda} - s} \right) = c_2$$

are two independent solutions. The general solution can be written in the form

$$G(s, t) = H \left[e^{(\lambda - \mu)t} \left(\frac{1 - s}{\frac{\mu}{\lambda} - s} \right) \right].$$

Here H is a function determined by the initial condition $G(s, 0) = s^{n_0}$. Find H and recover formula (6.22) for the probability generating function.

6.17. Apply Cauchy's method outlined in Problem 6.16 to the immigration model in Example 6.3. In this application the probability generating function satisfies

$$\frac{\partial G(s, t)}{\partial t} = \lambda(s - 1)G(s, t) + \mu(1 - s)\frac{\partial G(s, t)}{\partial s}.$$

Solve the equation assuming an initial population of n_0.

6.18. In a population sustained by immigration at rate λ with a simple death process with rate μ, the probability $p_n(t)$ satisfies

$$\frac{dp_0(t)}{dt} = -\lambda p_0(t) + \mu p_1(t),$$

$$\frac{dp_n(t)}{dt} = \lambda p_{n-1}(t) - (\lambda + n\mu)p_n(t) + (n + 1)\mu p_{n+1}(t).$$

Investigate the steady-state behaviour of the system by assuming that

$$p_n(t) \to p_n, \quad dp_n(t)/dt \to 0$$

for all n, as $t \to \infty$. Show that the resulting difference equations for what is known as the corresponding *stationary process*

$$-\lambda p_0 + \mu p_1 = 0,$$

$$\lambda p_{n-1} - (\lambda + n\mu)p_n + (n + 1)\mu p_{n+1} = 0, \qquad (n = 1, 2, \ldots)$$

can be solved iteratively to give

$$p_1 = \frac{\lambda}{\mu}p_0, \quad p_2 = \frac{\lambda^2}{2!\mu^2}p_0, \quad \cdots \quad p_n = \frac{\lambda^n}{n!\mu^n}p_0, \quad \cdots.$$

Using the condition $\sum_{n=0}^{\infty} p_n = 1$, and assuming that $\lambda < \mu$, determine p_0. Find the mean steady-state population size, and compare the result with that obtained in Example 6.3.

6.19. In a simple birth process the probability that the population is of size n at time t given that it was n_0 at time $t = 0$ is given by

$$p_n(t) = \binom{n-1}{n_0-1} e^{-\lambda n_0 t}(1 - e^{-\lambda t})^{n-n_0}, \qquad (n \geq n_0)$$

(see Section 6.3 and Figure 6.1). Show that the probability achieves its maximum value for given n and n_0 when $t = (1/\lambda) \ln(n/n_0)$. Find also the maximum value of $p_n(t)$ at this time.

6.20. In a birth and death process with equal birth and death parameters λ, the probability generating function is (see Eqn (6.24))

$$G(s,t) = \left[\frac{1 + (\lambda t - 1)(1 - s)}{1 + \lambda t(1 - s)} \right]^{n_0}.$$

Find the mean population size at time t. Show also that variance of the population size is $2n_0 \lambda t$.

6.21. In a death process the probability that a death occurs in time δt is the time-dependent parameter $\mu(t)n$ when the population size is n. The pgf $G(s,t)$ satisfies

$$\frac{\partial G}{\partial t} = \mu(t)(1 - s)\frac{\partial G}{\partial s}$$

as in Section 6.4. Show that

$$G(s,t) = [1 - e^{-\tau}(1 - s)]^{n_0},$$

where

$$\tau = \int_0^t \mu(x)\,dx.$$

Find the mean population size at time t.

In a death process it is found that the expected value of the population size at time t is given by

$$\mu(t) = \frac{n_0}{1 + \alpha t}, \qquad (t \geq 0),$$

where α is a positive constant. Find the corresponding death rate $\mu(t)$.

6.22. A population process has a probability generating function $G(s,t)$ which satisfies the equation

$$e^{-t}\frac{\partial G}{\partial t} = \lambda(s - 1)^2\frac{\partial G}{\partial s}.$$

If, at time $t = 0$, the population size is n_0, show that

$$G(s,t) = \left[\frac{1 + (1 - s)(\lambda e^t - \lambda - 1)}{1 + \lambda(1 - s)(e^t - 1)} \right]^{n_0}.$$

Find the mean population size at time t, and the probability of ultimate extinction.

6.23. A population process has a probability generating function given by

$$G(s,t) = \frac{1 - \mu e^{-t}(1 - s)}{1 + \mu e^{-t}(1 - s)},$$

where μ is a parameter. Find the mean of the population size at time t, and its limit as $t \to \infty$. Expand $G(s,t)$ in powers of s, and determine the probability that the population size is n at time t.

6.24. In a birth and death process with equal rates λ, the probability generating function is given by (see Eqn (6.24))

$$G(s,t) = \left[\frac{\lambda(z + t) - 1}{\lambda(z + t)} \right]^{n_0} = \left[\frac{1 + (\lambda t - 1)(1 - s)}{1 + \lambda t(1 - s)} \right]^{n_0},$$

where n_0 is the initial population size. Show that p_i, the probability that the population size is i at time t, is given by

$$p_i(t) = \sum_{m=0}^{i} \binom{n_0}{m} \binom{n_0 + i - m - 1}{i - m} \alpha(t)^m \beta(t)^{n_0 + i - m}$$

if $i \le n_0$, and by

$$p_i(t) = \sum_{m=0}^{n_0} \binom{n_0}{m} \binom{n_0 + i - m - 1}{i - m} \alpha(t)^m \beta(t)^{n_0 + i - m}$$

if $i > n_0$, where

$$\alpha(t) = \frac{1 - \lambda t}{\lambda t}, \qquad \beta(t) = \frac{\lambda t}{1 + \lambda t}.$$

6.25. We can view the birth and death process by an alternative differencing method. Let $p_{ij}(t)$ be the conditional probability

$$p_{ij}(t) = \mathbf{P}(N(t) = j | N(0) = i),$$

where $N(t)$ is the random variable representing the population size at time t. Assume that the process is in the (fixed) state $N(t) = j$ at times t and $t + \delta t$ and decide how this can arise from an incremental change δt in the time. If the birth and death rates are λ_j and μ_j, explain why

$$p_{ij}(t + \delta t) = p_{ij}(t)(1 - \lambda_i \delta t - \mu_i \delta t) + \lambda_i \delta t p_{i+1,j}(t) + \mu_i \delta t p_{i-1,j}(t) + o(\delta t)$$

for $i - 1, 2, 3, \ldots, j - 0, 1, 2, \ldots$. Take the limit as $\delta t \to 0$, and confirm that $p_{ij}(t)$ satisfies the differential equation

$$\frac{dp_{ij}(t)}{t} = -(\lambda_i + \mu_i) p_{ij}(t) + \lambda_i p_{i+1,j}(t) + \mu_i p_{i-1,j}(t).$$

How should $p_{0,j}(t)$ be interpreted?

6.26. Consider a birth and death process in which the rates are $\lambda_i = \lambda i$ and $\mu_i = \mu i$, and the initial population size is $n_0 = 1$. If

$$p_{1,j} = \mathbf{P}(N(t) = j | N(0) = 1),$$

it was shown in Problem 6.25 that $p_{1,j}$ satisfies

$$\frac{dp_{1,j}(t)}{dt} = -(\lambda + \mu) p_{1,j}(t) + \lambda p_{2,j}(t) + \mu p_{0,j}(t), \quad (j = 0, 1, 2, \ldots),$$

where

$$p_{0,j}(t) = \begin{cases} 0, & j > 0 \\ 1, & j = 0. \end{cases}$$

If

$$G_i(s, t) = \sum_{j=0}^{\infty} p_{ij}(t) s^j,$$

show that

$$\frac{\partial G_1(s, t)}{\partial t} = -(\lambda + \mu) G_1(s, t) + \lambda G_2(s, t) + \mu.$$

Explain why $G_2(s,t) = [G_1(s,t)]^2$ (see Section 6.5). Now solve what is effectively an ordinary differential equation for $G_1(s,t)$, and confirm that

$$G_1(s,t) = \frac{\mu(1-s) - (\mu - \lambda s)e^{-(\lambda - \mu)t}}{\lambda(1-s) - (\mu - \lambda s)e^{-(\lambda - \mu)t}},$$

as in Eqn (6.23) with $n_0 = 1$.

6.27. In a birth and death process with parameters λ and μ, $(\mu > \lambda)$, and initial population size n_0, show that the mean of time to extinction T_{n_0} is given by

$$\mathbf{E}(T_{n_0}) = n_0 \mu (\mu - \lambda)^2 \int_0^\infty \frac{te^{-(\mu - \lambda)t}[\mu - \mu e^{-(\mu - \lambda)t}]^{n_0 - 1}}{[\mu - \lambda e^{-(\mu - \lambda)t}]^{n_0 + 1}} dt.$$

If $n_0 = 1$, using integration by parts, evaluate the integral over the interval $(0, \tau)$, and then let $\tau \to \infty$ to show that

$$\mathbf{E}(T_1) = -\frac{1}{\lambda} \ln\left(\frac{\mu - \lambda}{\mu}\right).$$

6.28. A death process (see Section 6.4) has a parameter μ and the initial population size is n_0. Its probability generating function is

$$G(s,t) = [1 - e^{-\mu t}(1 - s)]^{n_0}.$$

Show that the mean time to extinction is

$$\frac{n_0}{\mu} \sum_{k=0}^{n_0 - 1} \frac{(-1)^k}{(k+1)^2} \binom{n_0 - 1}{k}.$$

6.29. A colony of cells grows from a single cell without deaths. The probability that a single cell divides into two cells in a time interval δt is $\lambda \delta t + o(\delta t)$. As in Problem 6.1, the probability generating function for the process is

$$G(s,t) = \frac{se^{-\lambda t}}{1 - (1 - e^{-\lambda t})s}.$$

By considering the probability

$$\mathbf{P}(T_n \le t) = F(t) = 1 - \sum_{k=1}^{n-1} p_k(t),$$

where T_n is the random variable representing the time that the population is of size $n (\ge 2)$ for the first time, show that

$$\mathbf{E}(T_n) = \frac{1}{\lambda} \sum_{k=1}^{n-1} \frac{1}{k}.$$

6.30. In a birth and death process, the population size represented by the random variable $N(t)$ grows as a simple birth process with parameter λ. No deaths occur until time T when the whole population dies. Suppose that the random variable T has an exponential distribution with parameter μ. The process starts with one individual at time $t = 0$. What is the probability that the population exists at time t, namely that $\mathbf{P}[N(t) > 0]$?

What is the conditional probability $\mathbf{P}[N(t) = n|N(t) > 0]$ for $n = 1, 2, \ldots$? Now show that

$$\mathbf{P}[N(t) = n] = e^{-(\lambda+\mu)t}(1 - e^{-\lambda t})^{n-1}.$$

Construct the probability generating function of this distribution, and find the mean population size at time t.

6.31. In a birth and death process, the variable birth and death rates are, for $t > 0$, respectively given by

$$\lambda_n(t) = \lambda(t)n > 0, \ (n = 0, 1, 2, \ldots) \quad \text{and} \quad \mu_n(t) = \mu(t)n > 0, \ (n = 1, 2, \ldots).$$

If $p_n(t)$ is the probability that the population size at time t is n, show that its probability generating function

$$G(s, t) = \sum_{n=0}^{\infty} p_n(t)s^n,$$

satisfies

$$\frac{\partial G}{\partial t} = (s - 1)[\lambda(t)s - \mu(t)]\frac{\partial G}{\partial s}.$$

Suppose that $\mu(t) = \alpha\lambda(t)$ $(\alpha > 0, \alpha \neq 1)$, and that the initial population size is n_0. Show that

$$G(s, t) = \left[\frac{1 - \alpha q(s, t)}{1 - q(s, t)}\right]^{n_0} \quad \text{where} \quad q(s, t) = \left(\frac{1 - s}{\alpha - s}\right)\exp\left[(1 - \alpha)\int_0^t \lambda(u)du\right].$$

Find the probability of extinction at time t.

6.32. A continuous time process has three states, E_1, E_2, and E_3. In time δt the probability of a change from E_1 to E_2 is $\lambda\delta t$, from E_2 to E_3 is also $\lambda\delta t$, and from E_2 to E_1 is $\mu\delta t$. E_3 can be viewed as an absorbing state. If $p_i(t)$ is the probability that the process is in state E_i $(i = 1, 2, 3)$ at time t, show that

$$p_1'(t) = -\lambda p_1(t) + \mu p_2(t), \quad p_2'(t) = \lambda p_1(t) - (\lambda + \mu)p_2(t), \quad p_3'(t) = \lambda p_2(t).$$

Find the probabilities $p_1(t)$, $p_2(t)$, $p_3(t)$, if the process starts in E_1 at $t = 0$.

The process survives as long as it is in states E_1 or E_2. What is the *survival probability*, that is, $\mathbf{P}(T \geq t)$, of the process?

6.33. In a birth and death process, the birth and death rates are given respectively by $\lambda(t)n$ and $\mu(t)n$ in Problem 6.31. Find the equation for the probability generating function $G(s, t)$. If $\boldsymbol{\mu}(t)$ is the mean population size at time t, show, by differentiating the equation for $G(s, t)$ with respect to s, that

$$\boldsymbol{\mu}'(t) = [\lambda(t) - \mu(t)]\boldsymbol{\mu}(t)$$

(assume that $(s - 1)\partial^2 G(s, t)/\partial s^2 = 0$ when $s = 1$). Then show that

$$\boldsymbol{\mu}(t) = n_0 \exp\left[\int_0^t [\lambda(u) - \mu(u)]du\right],$$

where n_0 is the initial population size.

CHAPTER 7

Queues

7.1 Introduction

Queues appear in many aspects of life. Some are clearly visible, as in the queues at supermarket check-out tills: others may be less obvious as in call-stacking at airline information telephone lines, or in hospital waiting lists. In the latter an individual waiting may have no idea how many persons are in front of him or her in the hospital appointments list. Generally we are interested in the long-term behaviour of queues for future planning purposes—does a queue increase with time or does it have a steady state, and if it does have a steady state, then, on average, how many individuals are there in the queue and what is the mean waiting time?

We can introduce some of the likely hypotheses behind queues by looking at a familiar example from banking. Consider a busy city centre cash dispensing machine (often called an ATM) outside a bank. As an observer, what do we see? Individuals or **customers** as they are known in queueing processes approach the dispenser or **server**. If there is no one using the machine, then the customer inserts a cash card, obtains cash, or transacts other business. There is a period of time when the customer is being served. On the other hand, if there is already someone at the till then the assumption is that the customer and succeeding ones form a queue and wait their turn.

For most queues the assumption is that customers are served in the order of arrival at the end of the queue. The basis of this is the '**first come, first served**' rule. However, if queueing is not taking place, then the next customer could be chosen at random from those waiting. This could be the case where the 'customers' are components in a manufacturing process which are being delivered, stored until required, and then chosen at random for the next process. In this application the arrivals could be regular and predictable. The simplest assumption with arrivals at a queue is that they have a Poisson distribution with parameter λ, so that λ is the average number of arrivals per unit time. From our discussion of Poisson process in Chapter 5, this implies that the probability of an arrival in the time interval $(t, t + \delta t)$ is $\lambda \delta t$ regardless of what happened before time t. And, of course, the probability of two or more arrivals in the time interval is negligible. As with the service time the density function of the time interval between arrivals is exponential:

$$\text{arrivals:} \qquad g(t) = \lambda e^{-\lambda t}, \qquad t \geq 0.$$

Generally, service times are difficult to model successfully.

We have to make assumptions about both the **service time** and the **customer arrivals**. We could assume that service takes a *fixed* time (discussed in Section 7.5), or more likely, that the service times are random variables which are independent for each customer. They could perhaps have a exponential density

$$\text{service:} \qquad f(t) = \mu e^{-\mu t}, \qquad t \geq 0.$$

This means that the customer spends on average time $1/\mu$ at the dispenser or being served.

In practice, however, there may be busy periods. Arrivals at our cash dispenser are unlikely to be uniformly distributed over a 24-hour operation: there will be more customers at peak periods during the day, and hence, in reality, an overall Poisson distribution does not hold. If a queue lengthens too much at peak times to the point of deterring customers, the bank might wish to install a second cash dispenser. Hence a queue with a **single server** now becomes a **two-server queue**. Generalising this, we can consider n-server queues, which are typical queueing processes that occur in supermarkets, the customer having to choose which check-out appears to have the quickest service time or shortest queues. Alternatively, the n servers could be approached through a single queue with customers being directed to the next free server, known as a **simple feeder queue**.

There are many aspects of queues which can be considered but not all of them will be investigated here. Customers could baulk at queueing if there are too many people ahead of them. Queues can have a maximum length: when the queue reaches a prescribed length, further customers are turned away.

Generally, we are not concerned with how queues begin when, say, a bank opens, but how they develop over time as the day progresses. Do the queues become longer, and if not what is the average length of the queue? Often, therefore, we look at the underlying **limiting** or **steady-state process** associated with the queue. We shall see that this effectively takes time variations out of the problem with the result that differential-difference equations for recurrent probabilities of the Markov process reduce to difference equations. Queueing processes are related to Poisson processes, and birth and death processes of Chapters 5 and 6 when they are Markov and the construction of the difference equations follows from similar arguments and hypotheses. In population terminology, joining a queue corresponds to a 'birth', and the end of the serving a 'death'.

7.2 The single-server queue

We will now put together a simple model of a queue. Let the random variable $N(t)$ denote the number of individuals in the queue *including the person being served*. We assume that there is a single counter or server with an orderly-forming queue served on a first come, first served basis. As before we write

$$\mathbf{P}(N(t) = n) = p_n(t),$$

which is the probability that there are n individuals in the queue at time t. We now look at the probability that there are n persons in the queue at time $t + \delta t$. We have to specify the probability that an individual arrives at the end of the queue in the time interval δt. Probably the simplest assumption is that the probability of an arrival is $\lambda \delta t$, where λ is a constant. In other words the arrivals form a Poisson process with intensity λ. As stated in the introduction we assume that these form random time intervals with exponential density given by

$$f(t) = \mu e^{-\mu t}, \qquad t \geq 0,$$

these times being measured from the arrival of a customer at the counter. The question is: what is the probability that a customer leaves the counter between the times t and $t + \delta t$? If T is a random variable representing the service time (the time spent at the counter by the customer), then

$$\mathbf{P}(t \leq T \leq t + \delta t | T \geq t)$$

represents the probability that the service will be completed in the interval $(t, t + \delta t)$, given that it is still in progress at time t. For this conditional probability (see Sections 1.3 and 1.5):

$$
\begin{aligned}
\mathbf{P}(t \leq T \leq t + \delta t | T \geq t) &= \frac{\mathbf{P}(t \leq T \leq t + \delta t \cap T \geq t)}{\mathbf{P}(T \geq t)} = \frac{\mathbf{P}(t \leq T \leq t + \delta t)}{\mathbf{P}(T \geq t)} \\
&= \frac{\int_t^{t+\delta t} \mu e^{-\mu s} ds}{\int_t^\infty \mu e^{-\mu s} ds} \\
&= \frac{[-e^{-\mu s}]_t^{t+\delta t}}{[-e^{-\mu s}]_t^\infty} = \frac{-e^{-\mu(t+\delta t)} + e^{-\mu t}}{e^{-\mu t}} \\
&= 1 - e^{-\mu \delta t} \approx 1 - (1 - \mu \delta t + o(\delta t)) \\
&= \mu \delta t + o(\delta t)
\end{aligned}
$$

for small δt, using a two-term series expansion for $e^{-\mu \delta t}$. As we would expect with this exponential distribution, the probability of the service being completed in the interval $(t, t + \delta t)$ is independent of the current state of the service—recall the no memory property.

By the law of total probability (Section 1.3) we have

$$p_n(t + \delta t) = \lambda \delta t p_{n-1}(t) + \mu \delta t p_{n+1}(t) + (1 - \lambda \delta t - \mu \delta t)p_n(t) + o(\delta t) \quad (n \geq 1),$$

$$p_0(t + \delta t) = \mu \delta t p_1(t) + (1 - \lambda \delta t)p_0(t),$$

where the probability of multiple events (arrivals or service) are assumed to be negligible. Thus

$$\frac{p_n(t + \delta t) - p_n(t)}{\delta t} = \lambda p_{n-1}(t) + \mu p_{n+1}(t) - (\lambda + \mu)p_n(t) + o(1),$$

$$\frac{p_0(t + \delta t) - p_0(t)}{\delta t} = \mu p_1(t) - \lambda p_0(t) + o(1).$$

Let $\delta t \to 0$, so that

$$
\left.
\begin{aligned}
\frac{dp_n(t)}{dt} &= \lambda p_{n-1}(t) + \mu p_{n+1}(t) - (\lambda + \mu) p_n(t), \qquad (n = 1, 2, \ldots) \\
\frac{dp_0(t)}{dt} &= \mu p_1(t) - \lambda p_0(t).
\end{aligned}
\right\} \quad (7.1)
$$

These equations are difficult to solve by the probability generating function method. Compared with birth and death equations of the previous chapter, whilst looking similar, the additional term $-\lambda p_(0)(t)$ in the second equation in (7.1) appears in the generating function equation (see Problem 7.7). However, it is possible to solve the equations using this method, and the result, after considerable analysis, leads to the formulas for probabilities in terms of special functions known as Bessel functions [1] (see Bailey (1964)). The partial differential equation for the probability generating function is derived in Problem 7.7. Fortunately we can obtain considerable information about queues by looking at the underlying **limiting process** in Eqn (7.1). Whilst time-dependent solutions are interesting, for planning purposes, the limiting process (if it exists) gives us a long-term view.

7.3 The limiting process

In the limiting process we look at the long-term behaviour of the queue assuming that it has a limiting state. Not all queues do: if rate of arrivals is large compared with the service time then we might expect the queue to continue to grow without bound as time progresses. For the moment let us assume that there is a limiting state, that $\lambda < \mu$, and that we can write

$$
\lim_{t \to \infty} p_n(t) = p_n.
$$

The probabilities now approach a *constant* distribution which does not depend on time. In this case it is reasonable to assume that

$$
\lim_{t \to \infty} \frac{dp_n(t)}{dt} = 0.
$$

If the limits turn out to be not justifiable then we might expect some inconsistency to occur in the limiting form of the equations or in their solution, which is a backward justification.

Equations (7.1) now become

$$
\left.
\begin{aligned}
\lambda p_{n-1} + \mu p_{n+1} - (\lambda + \mu) p_n &= 0, \qquad (n = 1, 2, \ldots) \\
\mu p_1 - \lambda p_0 &= 0.
\end{aligned}
\right\} \quad (7.2)
$$

The general equation in (7.2) is a second-order **difference equation**. It is easy to solve iteratively since

$$
p_1 = \frac{\lambda}{\mu} p_0
$$

---

[1] Friedrich Bessel (1784–1848), German mathematician.

and, with $n = 1$,

$$p_2 = \frac{1}{\mu}[(\lambda + \mu)p_1 - \lambda p_0] = \frac{1}{\mu}[(\lambda + \mu)\frac{\lambda}{\mu}p_0 - \lambda p_0] = \left(\frac{\lambda}{\mu}\right)^2 p_0.$$

The next iterate is

$$p_3 = \left(\frac{\lambda}{\mu}\right)^3 p_0,$$

and, in general,

$$p_n = \left(\frac{\lambda}{\mu}\right)^n p_0. \tag{7.3}$$

Alternatively the second-order difference equation can be solved using the characteristic equation method (see Section 2.2). The characteristic equation of (7.2) is

$$\mu m^2 - (\lambda + \mu)m + \lambda = 0, \quad \text{or} \quad (\mu m - \lambda)(m - 1) = 0.$$

The roots are $m_1 = 1$, $m_2 = \lambda/\mu$. If $\lambda \neq \mu$, then the general solution is

$$p_n = A + B\left(\frac{\lambda}{\mu}\right)^n; \tag{7.4}$$

if $\lambda = \mu$, it is

$$p_n = C + Dn,$$

where A, B, C, and D are constants. Since $\mu p_1 = \lambda p_0$, it follows that $A = 0$ or $D = 0$, and that $B = C = p_0$: the result is (7.3) again.

Let $\rho = \lambda/\mu$ for future reference: ρ is known as the **traffic density**. Generally, the traffic density is defined as the ratio of expected values:

$$\rho = \frac{\mathbf{E}(S)}{\mathbf{E}(I)},$$

where S and I are independent random variables of the service and inter-arrival times, respectively. When S and I are both exponentially distributed, then $\mathbf{E}(S) = 1/\mu$ and $\mathbf{E}(I) = 1/\lambda$ (Section 1.8). We still require the value of p_0. However, since $\sum_{n=0}^{\infty} p_n = 1$, it follows that

$$\left(\sum_{n=0}^{\infty} \rho^n\right) p_0 = 1,$$

but the geometric series (see Appendix) on the left will only *converge* if $|\rho| < 1$. Its sum is then given by

$$\sum_{n=0}^{\infty} \rho^n = \frac{1}{1 - \rho}.$$

Hence,

$$p_0 = 1 - \rho,$$

and

$$p_n = (1 - \rho)\rho^n, \quad (n = 0, 1, 2, \ldots), \tag{7.5}$$

which is the probability mass function of a geometric distribution with parameter $1 - \rho$ (see Section 1.7).

If $\rho \geq 1$ then the series does not converge and $\{p_n\}$ cannot represent a probability distribution. In this case the queue simply increases in length to infinity as $t \to \infty$ whatever its initial length. On the other hand if $\rho < 1$, then $\lambda < \mu$ and the rate of arrivals is less than the average service time with the result that a steady state should be achieved as we might expect. Notice that, if $\rho = 1$, the queue grows in length although there appears to be 'equilibrium' between arrivals and service.

There are various items of further information about the single-server queue, which we can deduce from the distribution $\{p_n\}$ assuming $\rho < 1$.

(i) **Server free.** The probability that the server is free when a customer arrives is

$$p_0 = 1 - \rho.$$

If the service time parameter ρ can be varied, for example, then the question might arise as to what might be an acceptable figure for p_0. If ρ cannot be significantly changed then perhaps more servers are required (see Section 7.4).

(ii) **Length of queue.** The mean length of the queue (including the person being served) is, if N represents a random variable of the number in the queue,

$$
\begin{aligned}
\mathbf{E}(N) &= \sum_{n=1}^{\infty} n p_n = \sum_{n=1}^{\infty} n(1 - \rho)\rho^n = \sum_{n=1}^{\infty} n\rho^n - \sum_{n=1}^{\infty} n\rho^{n+1} \\
&= \sum_{n=1}^{\infty} n\rho^n - \sum_{n=2}^{\infty} (n - 1)\rho^n = \sum_{n=1}^{\infty} \rho^n = \frac{\rho}{1 - \rho},
\end{aligned}
\tag{7.6}
$$

or the result could be quoted directly from the mean of the geometric distribution. Hence if $\rho = \frac{3}{4}$, then a customer might expect one person being served and two queueing. If, for example, the service time parameter μ can be varied (customers are served more efficiently, perhaps), then $\mathbf{E}(N) = 2$ could be set at a level acceptable to arriving customers. For example setting $\mathbf{E}(N) = 2$ implies that $\rho = \frac{2}{3}$, or $\mu = \frac{3}{2}\lambda$.

(iii) **Waiting time.** How long will a customer expect to wait for service on arrival at the back of the queue, and how long will the customer expect to queue and be served? A customer arrives and finds that there are n individuals ahead including the person being served. Let T_i be the random variable representing the time for the service of customer i. We are interested in the random variable

$$S_n = T_1 + T_2 + \cdots + T_n,$$

the sum of the random variables of service times of the first n queueing customers. These random variables are independent, each with an exponential density function with the same parameter μ, that is,

$$f(t) = \mu e^{-\mu t} \quad t \geq 0.$$

The moment generating function (see Section 1.9) of the random variable T_i is

$$
\begin{aligned}
M_{T_i}(s) &= \mathbf{E}(e^{sT_i}) = \int_0^\infty e^{su} f(u)du = \mu \int_0^\infty e^{su} e^{-\mu u} du, \\
&= \mu \int_0^\infty e^{-(\mu-s)u} du, \\
&= \frac{\mu}{\mu - s}, \qquad (s < \mu).
\end{aligned}
$$

We need the result that the moment generating function of the sum

$$
S_n = T_1 + T_2 + \cdots + T_n
$$

is

$$
\begin{aligned}
M_{S_n}(s) &= \mathbf{E}(e^{sS_n}) = \mathbf{E}(e^{s(T_1+T_2+\cdots+T_n)}) \\
&= \mathbf{E}(e^{sT_1})\mathbf{E}(e^{sT_2}) \cdots \mathbf{E}(e^{sT_n}) = M_{T_1}(s)M_{T_2}(s) \cdots M_{T_n}(s) \\
&= \left(\frac{\mu}{\mu - s}\right)^n
\end{aligned}
$$

(see Section 1.9). This is the moment generating function of the gamma density function with parameters μ and n. Its density is

$$
f_{S_n}(t) = \frac{\mu^n}{(n-1)!} t^{n-1} e^{-\mu t}
$$

(see Section 1.8). (Alternatively, the sum of *iid* exponential random variables is gamma with parameters μ and n.)

Let S be the random time a customer has to wait to reach the server, and as before, N is the random variable of the number in the queue. Using the law of total probability, the probability that S is greater than t is given by

$$
\mathbf{P}(S > t) = \sum_{n=1}^\infty \mathbf{P}(S_n > t | N = n)p_n.
$$

Hence,

$$
\begin{aligned}
\mathbf{P}(S > t) &= \sum_{n=1}^\infty p_n \int_t^\infty f_{S_n}(s)ds, \\
&= \sum_{n=1}^\infty (1-\rho)\rho^n \int_t^\infty \frac{\mu^n}{(n-1)!} s^{n-1} e^{-\mu s} ds, \\
&= (1-\rho)\mu\rho \int_t^\infty e^{-\mu s} \sum_{n=0}^\infty \frac{(\mu\rho s)^n}{n!} ds, \\
&= (1-\rho)\mu\rho \int_t^\infty e^{-\mu(1-\rho)s} ds, \\
&= \rho e^{-\mu(1-\rho)t}.
\end{aligned}
$$

The associated density function is

$$g(t) = \frac{d}{dt}[1 - \rho e^{-\mu(1-\rho)t}] = \rho\mu(1 - \rho)e^{-\mu(1-\rho)t}, \quad t \geq 0. \tag{7.7}$$

Finally, the expected value of S is

$$\begin{aligned}
\mathbf{E}(S) &= \int_0^\infty tg(t)dt = \int_0^\infty t\rho\mu(1 - \rho)e^{-\mu(1-\rho)t}dt, \\
&= \rho\mu(1 - \rho)\cdot\frac{1}{\mu^2(1 - \rho)^2}\int_0^\infty se^{-s}ds, \quad [s = \mu(1 - \rho)t], \\
&= \frac{\rho}{\mu(1 - \rho)}, \quad [\text{(see the Appendix for this integral)}], \tag{7.8}
\end{aligned}$$

which turns out to be $\mathbf{E}(T)$, the expected value $1/\mu$ of the service time T multiplied by the expected length of the queue $\rho/(1 - \rho)$ from (ii) above. Hence, in this case, it occurs that

$$\mathbf{E}(S) = \mathbf{E}(N)\mathbf{E}(T).$$

To use this result directly we would need to show that N and T are independent random variables. With n customers ahead the next customer could expect to wait for a time $\rho/(\mu(1 - \rho))$ to reach the server. Until service is completed the customer could expect to spend time

$$\frac{\rho}{\mu(1 - \rho)} + \frac{1}{\mu} = \frac{1}{\mu(1 - \rho)}$$

since the expected service time is $1/\mu$.

This is a lengthy argument: the difficulty is mainly caused by the problem of finding the density function of the sum of a set of random variables.

(iv) **Busy periods.** If $N(t)$ is the random number representing the number of individuals in a single-server queue at time t in the time-dependent case (Eqn (7.1)), then the development of a queue over time with $\rho = \lambda/\mu < 1$ might look as in Figure 7.1. This is a **realisation** (or **sample path**) of a continuous time process which has discrete jumps at varying times. There will be periods where the server is free (called **slack periods**), periods of lengths

$$s_1 = t_1, \quad s_2 = t_3 - t_2, \quad s_3 = t_5 - t_4, \ldots,$$

and periods when the server is busy, namely

$$b_1 = t_2 - t_1, \quad b_2 = t_4 - t_3, \quad b_3 = t_6 - t_5, \ldots.$$

The times $t_1, t_3, t_5, \ldots$ are the times when a new customer arrives when the server is free, and $t_2, t_4, t_6, \ldots$ are the times at which the server becomes free. The periods denoted by $b_1, b_2, \ldots$ are known as **busy periods**. A question whose answer is of interest is: what is the expected length of a busy period? Here we present an informal argument which points to the result.

Suppose that the server is free at time $t = t_2$. Then since the arrivals form a Poisson process with parameter λ, the expected time until the next customer arrives

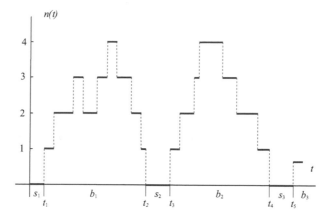

Figure 7.1 *A sample queue for the number in the queue $N(t)$ against time in the time-dependent case: the queue starts with no customers and $\{s_i\}$ are the slack periods and $\{b_i\}$ are the busy periods.*

is $1/\lambda$, since the density function is $\lambda e^{-\lambda t}$. It seems a reasonable intuitive result that the average lengths of a large number n of slack periods should approach $1/\lambda$, that is,

$$\lim_{n \to \infty} \left[\frac{\sum_{i=1}^{n} s_i}{n} \right] = \frac{1}{\lambda}.$$

In the limiting process, the probability that the server is free is p_0 and that the server is busy is $1 - p_0$. Hence the ratio of the average lengths of the slack periods to that of the busy periods is $p_0/(1 - p_0)$. Hence

$$\lim_{n \to \infty} \left[\frac{\sum_{i=1}^{n} s_i}{n} \cdot \frac{n}{\sum_{i=1}^{n} b_i} \right] - \frac{p_0}{1 - p_0} = \frac{1 - \rho}{\rho}.$$

Assuming that the limit of the product is the product of the limits, it follows using the limit above that

$$\lim_{n \to \infty} \left[\frac{1}{n} \sum_{i=1}^{n} b_i \right] = \frac{1}{\lambda} \frac{\rho}{1 - \rho} = \frac{1}{\mu - \lambda}, \tag{7.9}$$

which is the mean length of the busy periods.

Example 7.1 (The baulked queue) *In this queue not more than $m \geq 2$ people (including the person being served) are allowed to form a queue. If there are m individuals in the queue, then any further arrivals are turned away. It is assumed that the service distribution is exponential with rate μ, and arrival distribution is exponential with parameter λ if $n < m$ where n is the number in the queue. Find the probability distribution for the queue length.*

If $n < m$ the arrival rate is λ, but if $n \geq m$ we assume that $\lambda = 0$. The difference equations in (7.2) are modified so that they become the finite system

$$\mu p_1 - \lambda p_0 = 0,$$

$$\lambda p_{n-1} + \mu p_{n+1} - (\lambda + \mu)p_n = 0, \qquad (1 \le n \le m - 1),$$

$$\lambda p_{m-1} - \mu p_m = 0.$$

The general solution is (see Eqn (7.4))

$$p_n = A + B\rho^n, \qquad (\rho \ne 1).$$

For the baulked queue there is no restriction on the size of $\rho = \lambda/\mu$ since the queue length is controlled The case $\rho = 1$ is treated separately.

The *boundary conditions* imply

$$\mu(A + B\rho) - \lambda(A + B) = 0, \quad \text{or}, \quad A(\mu - \lambda) = 0.$$

Hence $A = 0, (\mu \ne \lambda)$. Therefore

$$p_n = B\rho^n, \qquad (n = 0, 1, 2, \ldots, m).$$

Unlike the difference equation in Section 2.2, there no second boundary condition. However, a further condition follows since $\{p_n(t)$ must be a probability distribution so that $\sum_{n=0}^{m} p_n = 1$. Hence

$$B \sum_{n=0}^{m} \rho^n = 1,$$

or, after summing the geometric series,

$$B \frac{1 - \rho^{m+1}}{1 - \rho} = 1.$$

Hence, the probability that there are n individuals in the queue is given by

$$p_n = \frac{\rho^n(1 - \rho)}{(1 - \rho^{m+1})}, \qquad (n = 0, 1, 2, \ldots, m) \quad (\rho \ne 1).$$

If $\rho = 1$, then the general solution for p_n is

$$p_n = A + Bn,$$

and the boundary condition $p_1 - p_0 = 0$ implies $B = 0$. Hence $p_n = A$ $(n = 0, 1, 2, \ldots m)$, but since $\sum_{n=0}^{m} p_n = 1$, it follows that $A = 1/(m + 1)$. Finally

$$p_n = \frac{1}{m + 1}, \qquad (n = 0, 1, 2, \ldots m),$$

a discrete uniform distribution which does not depend on n.

7.4 Queues with multiple servers

In many practical applications of queueing models, there is more than one server, as in banks, supermarkets, and hospital admissions. Usually in supermarkets, shoppers at check-outs choose a till with the shortest queue or, more accurately, a queue with the smallest number of likely purchases. On the other, hand banks and post offices, for example, often guide arrivals into a single queue, and then direct them to tills or counters as they become free. It is the second type of queue which we will investigate. A plan of such a scheme is shown in Figure 7.2.

Suppose that the queue faces r servers. As for the single-server queue, assume that the arrivals form a Poisson process with parameter λ, and that the service time at each counter has exponential distribution with rate μ. Let $p_n(t)$ be the probability

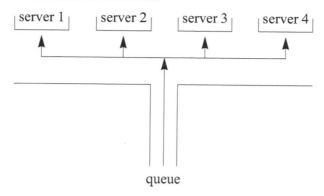

Figure 7.2 *A single queue with 4 servers.*

that there are n people in the queue at time t including those being served. The queue length is governed by different equations depending on whether $n < r$ or $n \geq r$. If $n \geq r$ then all counters will be occupied by a customer, but some will be free if $n < r$. As before, if $n < r$ the probability of a counter becoming free in time δt will be $n\mu\delta t$: if $n \geq r$ the probability will be $r\mu\delta t$. Modifying the arguments of Section 7.2 we can derive the following incremental equations:

$$p_0(t + \delta t) = \mu p_1(t)\delta t + (1 - \lambda)p_0(t)\delta t + o(\delta t),$$

$$p_n(t + \delta t) = \lambda p_{n-1}(t)\delta t + (n + 1)\mu p_{n+1}(t)\delta t + (1 - \lambda - n\mu)p_n(t)\delta t + o(\delta t),$$
$$(1 < n < r).$$

$$p_n(t + \delta t) = \lambda p_{n-1}(t)\delta t + r\mu p_{n+1}(t)\delta t + (1 - \lambda - r\mu)p_n(t)\delta t + o(\delta t),$$
$$(n \geq r).$$

With the usual limiting process we obtain the time-dependent equations for the queueing process:

$$\frac{dp_0(t)}{dt} = \mu p_1(t) - \lambda p_0(t),$$

$$\frac{dp_n(t)}{dt} = \lambda p_{n-1}(t) + (n + 1)\mu p_{n+1}(t) - (\lambda + n\mu)p_n(t), \qquad (1 \leq n < r),$$

$$\frac{dp_n(t)}{dt} = \lambda p_{n-1}(t) + r\mu p_{n+1}(t) - (\lambda + r\mu)p_n(t), \qquad (n \geq r).$$

Assuming that the derivatives tend to zero as $t \to \infty$ (as in the previous section), the corresponding limiting process (if it exists) for the queue with r servers is

$$\mu p_1 - \lambda p_0 = 0, \tag{7.10}$$

$$\lambda p_{n-1} + (n + 1)\mu p_{n+1} - (\lambda + n\mu)p_n = 0, \qquad 1 \leq n < r, \tag{7.11}$$

$$\lambda p_{n-1} + r\mu p_{n+1} - (\lambda + r\mu)p_n = 0, \qquad n \geq r. \tag{7.12}$$

This is not now a set of constant-coefficient difference equations, and the characteristic equation method of solution no longer works. However, it is easy to solve the equations iteratively. For $n = 1$ in Eqn (7.11), we have

$$\lambda p_0 + 2\mu p_2 - (\lambda + \mu)p_1 = 0,$$

or, using Eqn (7.10),

$$2\mu p_2 - \lambda p_1 = 0.$$

Provided $n < r$ we can repeat this procedure for $n = 2$ giving

$$3\mu p_3 = (\lambda + 2\mu)p_2 - \lambda p_1 = (\lambda + 2\mu)p_2 - 2\mu p_2 = \lambda p_2,$$

and so on:

$$4\mu p_4 = \lambda p_3, \ \ldots, \ r\mu p_r = \lambda p_{r-1}.$$

For $n \geq r$, we switch to the difference equations (7.8) and the sequence then continues as

$$r\mu p_{r+1} = (\lambda + r\mu)p_r - \lambda p_{r-1} = \lambda p_r, \ \ldots, \ r\mu p_n = \lambda p_{n-1}, \ \ldots$$

etc. To summarise: the set of difference equations (7.10)–(7.12) reduce to the equivalent set

$$n\mu p_n = \lambda p_{n-1}, \qquad (1 \leq n < r), \tag{7.13}$$

$$r\mu p_n = \lambda p_{n-1}, \qquad (n \geq r). \tag{7.14}$$

Starting with $n = 1$,

$$p_1 = \frac{\lambda}{\mu}p_0 = \rho p_0,$$

where $\rho = \lambda/\mu$. Then for $n = 2$,

$$p_2 = \frac{\rho}{2}p_1 = \frac{\rho^2}{2}p_0,$$

and so on. Thus for $n < r$,

$$p_n = \frac{\rho^n}{n!}p_0,$$

and for $n \geq r$

$$p_n = \frac{\rho^n}{r^{n-r}r!}p_0.$$

Since $\sum_{n=0}^{\infty} p_n = 1$, we can determine p_0 from

$$p_0 \left[\sum_{n=0}^{r-1} \frac{\rho^n}{n!} + \frac{r^r}{r!} \sum_{n=r}^{\infty} \left(\frac{\rho}{r}\right)^n \right] = 1,$$

in which the infinite geometric series converges if $\rho < r$. As we would expect in the multiple-server queue, it is possible for the arrival rate λ to be greater than the service rate μ. Summing the second geometric series, it gives

$$\sum_{n=r}^{\infty} \left(\frac{\rho}{r}\right)^n = \frac{r}{r-\rho} \left(\frac{\rho}{r}\right)^r$$

so that

$$p_0 = 1 \left/ \left[\sum_{n=0}^{r-1} \frac{\rho^n}{n!} + \frac{\rho^r}{(r-\rho)(r-1)!} \right] \right. . \tag{7.15}$$

If N is the random variable of the queue length n, then the expected queue length *excluding* those being served is

$$\mathbf{E}(N) = \sum_{n=r+1}^{\infty} (n-r)p_n,$$

$$= \frac{p_0 \rho^r}{r!} \sum_{n=r+1}^{\infty} (n-r) \left(\frac{\rho}{r}\right)^{n-r}.$$

We need to sum the series. Let

$$R = \sum_{n=r+1}^{\infty} (n-r) \left(\frac{\rho}{r}\right)^{n-r} = \frac{\rho}{r} + 2\left(\frac{\rho}{r}\right)^2 + 3\left(\frac{\rho}{r}\right)^3 + \cdots.$$

Multiply both sides by ρ/r and subtract the new series from the series for R. Hence,

$$R\left(1 - \frac{\rho}{r}\right) = \left(\frac{\rho}{r}\right) + \left(\frac{\rho}{r}\right)^2 + \left(\frac{\rho}{r}\right)^3 + \cdots = \frac{\rho}{r - \rho},$$

using the formula for the sum of a geometric series. Hence

$$R = \frac{\rho r}{(r - \rho)^2},$$

and, therefore

$$\mathbf{E}(N) = \frac{p_0 \rho^{r+1}}{(r-1)!(r-\rho)^2}, \tag{7.16}$$

where p_0 is given by Eqn (7.15).

This formula gives the expected length of those actually waiting for service. The expected length including those being served can be found in the solution to Problem 7.10.

For a queue with r servers, $\rho/r = \lambda/(r\mu)$ is a measure of the traffic density of the queue, and it is this parameter which must be less than 1 for the expected length of the queue to remain finite. If, for example, $\rho = \lambda/\mu = 4$, then at least 5 servers are required to stop queue growth. For $r = 5$,

$$p_0 = 1 \Big/ \left[\sum_{n=0}^{4} \frac{4^n}{n!} + \frac{4^5}{4!}\right] = \frac{1}{77} = 0.013\ldots,$$

and

$$\mathbf{E}(N) = \frac{p_0 4^6}{4!} = \frac{512}{231} = 2.216\ldots.$$

The number of persons waiting averages about 2.22. Adding an extra server reduces the expected length of the waiting queue to $\mathbf{E}(N) = 0.57$.

This type of analysis could be used to determine the number of servers given estimates of λ and μ.

Example 7.2. *A bank has two tellers who take different mean service times to complete the service required by a customer. The customers arrive as a Poisson process with intensity λ, and the service times are independent and exponentially distributed with rates μ_1 and μ_2. If* *both tellers are free the tellers are equally likely to be chosen. In the limiting case, find the difference equations for the probability p_n that there are n persons in the queue including those being served. Solve the equations for p_n. What is the expected length of the queue?*

Since customers express no preference of tellers, by comparison with Eqns (7.10)–(7.12), the difference equations for the limiting process are

$$\tfrac{1}{2}(\mu_1 + \mu_2)p_1 - \lambda p_0 = 0,$$

$$(\mu_1 + \mu_2)p_2 + \lambda p_0 - [\lambda + \tfrac{1}{2}(\mu_1 + \mu_2)]p_1 = 0,$$

$$(\mu_1 + \mu_2)p_{n+1} + \lambda p_{n-1} - [\lambda + (\mu_1 + \mu_2)]p_n = 0, \qquad n > 2.$$

If we let $\mu = \tfrac{1}{2}(\mu_1 + \mu_2)$, then this is equivalent to the two-server queue, each server having exponential distribution with the same rate μ. Hence the required answers can be read off from the results for the multi-server queue presented immediately before this example. Thus, provided $\lambda < 2\mu$,

$$p_1 = \rho p_0,$$

$$p_n = \frac{\rho^n}{2^{n-2}2!}p_0, \qquad n \geq 2,$$

where $\rho = \lambda/\mu$, and

$$p_0 = 1 \left/ \left[1 + \rho + \frac{\rho^2}{2 - \rho}\right] = \frac{2 - \rho}{2 + \rho}.\right.$$

In this case the expected length of the queue is (excluding those being served)

$$\mathbf{E}(N) = \frac{p_0\rho^3}{(2 - \rho)^2} = \frac{\rho^3}{4 - \rho^2}.$$

Example 7.3. *Customers arrive in a bank with two counters at a rate $2\lambda\delta t$ in any small time interval δt. Service times at either counter is exponentially distributed with rate μ. Which of the following schemes leads to the shorter overall queue length?*
(a) A single queue feeding two servers.
(b) Two separate single-server queues with the assumptions that customers arrive at each queue with parameter $\lambda\delta t$, and choose servers at random.

The answer is reasonably obvious but it is useful to compare the outcomes.
(a) This is the two-server queue with rate 2λ. Hence, from (7.15) with $r = 2$,

$$p_0 = \frac{2 - \rho_1}{(1 + \rho_1)(2 - \rho_1) + \rho_1^2} = \frac{2 - \rho_1}{2 + \rho_1} = \frac{1 - \rho}{1 + \rho},$$

where $\rho_1 = 2\rho = 2\lambda/\mu$. From (7.16), the expected queue length will be, except for those being served,

$$\mathbf{E}_a(N) = \frac{p_0\rho_1^3}{(2 - \rho_1)^2} = \frac{\rho_1^3}{(4 - \rho_1^2)} = \frac{(2\lambda/\mu)^3}{4 - (2\lambda/\mu)^2} = \frac{2\rho^3}{1 - \rho^2}, \qquad (7.17)$$

where $\rho = \lambda/\mu$.
(b) Note that customers do not exert a choice in this queueing scheme: it is not quite the same problem as customers arriving and choosing the shorter queue. For the scheme as described, the expected length of the two single-server queues is twice that of one, and is, excluding those being served,

$$\mathbf{E}_b(N) = 2\sum_{n=2}^{\infty}(n - 1)p_n,$$

where $p_n = (1 - \rho)\rho^n$ from (7.5). Hence

$$\mathbf{E}_b(N) = 2 \left[\frac{\rho}{1 - \rho} - p_1 - (1 - p_0 - p_1) \right]$$

using the results

$$\sum_{n=1}^{\infty} n p_n = \frac{\rho}{1 - \rho}, \qquad \sum_{n=0}^{\infty} p_n = 1.$$

Hence, since $p_0 = 1 - \rho$ and $p_1 = (1 - \rho)\rho$,

$$\mathbf{E}_b(N) = \frac{2\rho^2}{1 - \rho}. \qquad (7.18)$$

We must compare $\mathbf{E}_a(N)$ and $\mathbf{E}_b(N)$ given by (7.13) and (7.14). Thus

$$\mathbf{E}_a(N) = \frac{2\rho^3}{1 - \rho^2} = \frac{\rho}{1 + \rho} \mathbf{E}_b(N) < \mathbf{E}_b(N)$$

for all ρ such that $0 < \rho < 1$. We conclude in this model that the two single-server queues have a longer expected total queue length than the two-server queue, which might perhaps be expected. Comparative graphs of $\mathbf{E}_a(N)$ and $\mathbf{E}_b(N)$ versus r are shown in Figure 7.3.

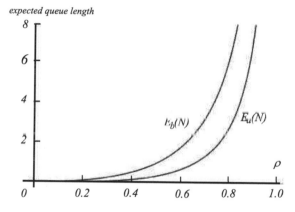

Figure 7.3 *Comparison of expected queue lengths for the two-server* $\mathbf{E}_a(N)$ *queue and the two single-server queues* $\mathbf{E}_b(N)$ *against {arrival rate}/{service rate} ratio.*

Again such an analysis could be useful in planning the type of queueing system to install.

7.5 Queues with fixed service times

In some queues the service time is a fixed value, say τ, since the customers all go through the same service. This could arise, for example, at a bank ATM where cash is dispensed. Suppose that the density function of the interval between arrivals is exponential with parameter λ, namely $\lambda e^{-\lambda t}$, that is, a Poisson arrival rate.

This queueing problem requires a different approach. As before, let $p_n(t)$ be the probability that there are n persons in the queue at time t including the person being served. We consider time intervals of duration τ, the service time, and start by asking: what is the probability $p_0(t+\tau)$ that the queue has no customers at time $t+\tau$? This can arise if there was either no one in the queue at time t, or one person at time t who must have been served and has left. The probability that this occurs is $p_0(t) + p_1(t)$. Also, there must have been no arrivals. Since the inter-arrival times have a negative exponential distribution, the probability of no arrivals in any time interval of duration τ is $e^{-\lambda\tau}$ (and remember its no-memory property). Hence

$$p_0(t+\tau) = [p_0(t) + p_1(t)]e^{-\lambda\tau}.$$

We now generalize this equation. Suppose that there are n persons in the queue at time $t + \tau$. This could have arisen in any of the ways listed below:

Number in queue at time t	Number of arrivals	Probability
0	n	$p_0(t)e^{-\lambda\tau}(\lambda\tau)^n/n!$
1	n	$p_1(t)e^{-\lambda\tau}(\lambda\tau)^n/n!$
2	$n-1$	$p_2(t)e^{-\lambda\tau}(\lambda\tau)^{n-1}/(n-1)!$
$\ldots$	$\ldots$	$\ldots$
n	1	$p_n(t)e^{-\lambda\tau}\lambda\tau/1!$
$n+1$	0	$p_{n+1}(t)e^{-\lambda\tau}$

By way of explanation, for example, in row 3 in the table, the one being served will leave at time $t + \tau$, which means that the number of arrivals must be $n - 1$: this occurs with probability that there are two in the queue at time t, namely $p_2(t)$ times $e^{-\lambda\tau}(\lambda\tau)^{n-1}/(n-1)!$. In general the probability of r arrivals in time τ is $e^{-\lambda\tau}(\lambda\tau)^r/r!$. Thus $p_n(t+\tau)$ is the sum of the probabilities in the third column:

$$p_n(t+\tau) = e^{-\lambda\tau}\left[p_0(t)\frac{(\lambda\tau)^n}{n!} + \sum_{r=1}^{n+1}\frac{p_r(t)(\lambda\tau)^{n+1-r}}{(n+1-r)!}\right] \quad (n = 0, 1, \ldots).$$

Suppose now that we just consider the long-term behavior of the queue by assuming that

$$\lim_{t\to\infty} p_n(t) = \lim_{t\to\infty} p_n(t+\tau) = p_n,$$

since τ is finite, and that the limits exist. Then the sequence $\{p_n\}$ satisfies the difference equations

$$p_n = e^{-\lambda\tau}\left[p_0\frac{(\lambda\tau)^n}{n!} + \sum_{r=1}^{n+1}\frac{p_r(\lambda\tau)^{n+1-r}}{(n+1-r)!}\right] \quad (n = 0, 1, \ldots).$$

For the next step it is more convenient to write the equations as a list, as follows:

$$\begin{aligned}
p_0 &= e^{-\lambda\tau}(p_0 + p_1) \\
p_1 &= e^{-\lambda\tau}[(p_0 + p_1)\lambda\tau + p_2] \\
p_2 &= e^{-\lambda\tau}\left[(p_0 + p_1)\frac{(\lambda\tau)^2}{2!} + p_2(\lambda\tau) + p_3\right]
\end{aligned}$$

$$\vdots \quad = \quad \vdots$$

$$p_n \quad = \quad e^{-\lambda\tau}\left[(p_0+p_1)\frac{(\lambda\tau)^n}{n!} + p_2\frac{(\lambda\tau)^{n-1}}{(n-1)!} + \cdots + p_n(\lambda\tau) + p_{n+1}\right]$$

$$\vdots \quad = \quad \vdots$$

We now construct the probability generating function for the sequence $\{p_n\}$ by multiplying p_n by s^n in the list above, and summing over n *by columns* to give

$$
\begin{aligned}
G(s) = \sum_{n=0}^{\infty} p_n s^n &= e^{-\lambda\tau}\left[(p_0+p_1)e^{\lambda\tau s} + p_2 s e^{\lambda\tau s} + p_3 s^2 e^{\lambda\tau s} + \cdots\right]\\
&= e^{-\lambda\tau(1-s)}\left[p_0 + p_1 + p_2 s + p_3 s^2 + \cdots\right]\\
&= e^{-\lambda\tau(1-s)}\left[p_0 + \frac{1}{s}\{G(s) - p_0\}\right].
\end{aligned}
$$

Hence, solving this equation for $G(s)$, we obtain

$$G(s) = \frac{p_0(1-s)}{1 - se^{\lambda\tau(1-s)}}.$$

However, this formula for $G(s)$ still contains an unknown constant p_0. For $G(s)$ to be a *pgf*, we must have $G(1) = 1$. Since both the numerator and denominator tend to zero as $s \to 1$, we must use an expansion to cancel a factor $(1-s)$. Let $u = 1-s$. Then

$$G(s) = H(u) = \frac{p_0 u}{1 - (1-u)e^{\lambda\tau u}}. \tag{7.19}$$

The Taylor series expansion of $H(u)$ about $u = 0$ is

$$H(u) = \frac{p_0}{1-\lambda\tau} - \frac{p_0\,\lambda\tau(2-\lambda\tau)}{2(1-\lambda\tau)^2}u + O(u^2)$$

as $u \to 0$. Hence

$$G(s) = \frac{p_0}{1-\lambda\tau} - \frac{p_0\lambda\tau(2-\lambda\tau)}{2(1-\lambda\tau)^2}(1-s) + O((1-s)^2) \tag{7.20}$$

as $s \to 1$. $G(1) = 1$ implies $p_0 = 1 - \lambda\tau$.

Individual probabilities can be found by expanding $G(s)$ in powers of s. We shall not discuss convergence in detail here, but simply state that the expansion converges for $0 \le \lambda\tau < 1$, a result to be expected intuitively.

The expected queue length is given by $G'(1)$ (see Section 1.9). From (7.19), it follows that

$$\mu = G'(1) = \frac{p_0\lambda\tau(2-\lambda\tau)}{2(1-\lambda\tau)^2} = \frac{\lambda\tau(2-\lambda\tau)}{2(1-\lambda\tau)},$$

since $p_0 = 1 - \lambda\tau$.

This can be expressed in terms of the traffic intensity ρ since $1/\lambda$ is the mean inter-arrival time and τ is the actual service time. Thus $\rho = \lambda\tau$, and the mean queue length is

$$\mu = \frac{\rho(1 - \frac{1}{2}\rho)}{1 - \rho}.$$

7.6 Classification of queues

There are three main factors which characterize queues: the probability distributions controlling arrivals, service times, and the number of servers. If the inter-arrival and service distributions are denoted by G_1 and G_2, and there are n servers, then this queue is described as a $G_1/G_2/n$ queue, which is shorthand for

arrival distribution, G_1/ service distribution, G_2 / number of servers, n.

If the inter-arrival and service distributions are exponentials with densities $\lambda e^{-\lambda t}$ and $\mu e^{-\mu t}$ (that is, both are Poisson processes), then both processes are Markov with parameters λ and μ, respectively. The Markov property means that the probability of the next arrival or the probability of service being completed are independent of any previous occurrences. We denote the processes by $M(\lambda)$ and $M(\mu)$ (M for Markov). Hence the corresponding single-server queue is denoted by $M(\lambda)/M(\mu)/1$, and the n-server queue by $M(\lambda)/M(\mu)/n$.

The single-server queue with Markov inter-arrival but fixed service time τ, discussed in the previous section, is denoted by $M(\lambda)/D(\tau)/1$, where D stands for deterministic.

This can be extended to other distributions. If the service time for a single-server queue has a uniform distribution with density function

$$f(t) = \begin{cases} \mu & 0 \leq t \leq 1/\mu, \\ 0 & \text{elsewhere,} \end{cases}$$

then this would be described as an $M(\lambda)/U(\mu)/1$ queue.

7.7 Problems

7.1. In a single-server queue, a Poisson process for arrivals of intensity $\frac{1}{2}\lambda$ and for service and departures of intensity λ are assumed. For the corresponding limiting process, find
   (a) p_n, the probability that there are n persons in the queue,
   (b) the expected length of the queue,
   (c) the probability that there are not more than two persons in the queue, including the person being served in each case.

7.2. Consider a telephone exchange with a very large number of lines available. If n lines are busy the probability that one of them will become free in small time δt is $n\mu\delta t$. The probability of a new call is $\lambda\delta t$ (that is, Poisson), with the assumption that the probability of multiple calls is negligible. Show that $p_n(t)$, the probability that n lines are busy at time t satisfies

$$p_0'(t) = -\lambda p_0(t) + \mu p_1(t),$$

$$p_n'(t) = -(\lambda + n\mu)p_n(t) + \lambda p_{n-1}(t) + (n+1)\mu p_{n+1}(t), \quad (n \geq 1).$$

In the limiting process, show by induction that

$$p_n = \lim_{t \to \infty} p_n(t) = \frac{e^{-\lambda/\mu}}{n!} \left(\frac{\lambda}{\mu}\right)^n.$$

Identify the distribution.

7.3. For a particular queue, when there are n customers in the system, the probability of an arrival in the small time interval δt is $\lambda_n \delta t + o(\delta t)$. The service time parameter μ_n is also a function of n. If p_n denotes the probability that there are n customers in the queue in the steady state queue, show by induction that

$$p_n = p_0 \frac{\lambda_0 \lambda_1 \dots \lambda_{n-1}}{\mu_1 \mu_2 \dots \mu_n}, \quad (n = 1, 2, \dots),$$

and find an expression for p_0.

If $\lambda_n = 1/(n+1)$ and $\mu_n = \mu$, a constant, find the expected length of the queue.

7.4. In a baulked queue (see Example 7.1), not more than $m \geq 2$ people are allowed to form a queue. If there are m individuals in the queue where m is fixed, then any further arrivals are turned away. If the arrivals form a Poisson process with intensity λ and the service distribution is exponential with parameter μ, show that the expected length of the queue is

$$\frac{\rho - (m+1)\rho^{m+1} + m\rho^{m+2}}{(1-\rho)(1-\rho^{m+1})},$$

where $\rho = \lambda/\mu \neq 1$. Deduce the expected length if $\rho = 1$. What is the expected length of the queue if $m = 3$ and $\rho = 1$?

7.5. Consider the single-server queue with Poisson arrivals occurring with intensity λ, and exponential service times with parameter μ. In the stationary process, the probability p_n that there are n individuals in the queue is given by

$$p_n = \left(1 - \frac{\lambda}{\mu}\right)\left(\frac{\lambda}{\mu}\right)^n, \quad (n = 0, 1, 2, \dots).$$

Find its probability generating function

$$G(s) = \sum_{n=0}^{\infty} p_n s^n.$$

Also find the mean and variance of the queue length.

7.6. A queue is observed to have an average length of 2.8 individuals including the person being served. Assuming the usual exponential distributions for both service times and times between arrivals, what is the traffic density, and the variance of the queue length?

7.7. The differential-difference equations for a queue with parameters λ and μ are (see Eqn (7.1))

$$\frac{dp_0(t)}{dt} = \mu p_1(t) - \lambda p_0(t),$$

$$\frac{dp_n(t)}{dt} = \lambda p_{n-1}(t) + \mu p_{n+1}(t) - (\lambda + \mu)p_n(t),$$

where $p_n(t)$ is the probability that the queue has length n at time t. Let the probability generating function of the distribution $\{p_n(t)\}$ be

$$G(s, t) = \sum_{n=0}^{\infty} p_n(t) s^n.$$

Show that $G(s,t)$ satisfies the equation

$$s\frac{\partial G(s,t)}{\partial t} = (s-1)(\lambda s - \mu)G(s,t) + \mu(s-1)p_0(t).$$

Unlike the birth and death processes in Chapter 6, this equation contains the unknown probability $p_0(t)$, which complicates its solution. Show that it can be eliminated to leave the following second-order partial differential equation for $G(s,t)$:

$$s(s-1)\frac{\partial^2 G(s,t)}{\partial t \partial s} - (s-1)^2(\lambda s - \mu)\frac{\partial G(s,t)}{\partial s} - \frac{\partial G(s,t)}{\partial t} - \lambda(s-1)^2 G(s,t) = 0.$$

This equation can be solved by Laplace[2] transform methods.

7.8. A call center has r telephones manned at any time, and the traffic density is $\lambda/(r\mu) = 0.86$. Compute how many telephones should be manned in order that the expected number of callers waiting at any time should not exceed 4. Assume a limiting process with inter-arrival times of calls and service times for all operators, both exponential with parameters λ and μ, respectively (see Section 7.4).

7.9. Compare the expected lengths of the two queues $M(\lambda)/M(\mu)/1$ and $M(\lambda)/D(1/\mu)/1$ with $\rho = \lambda/\mu < 1$. The queues have parameters such that the mean service time for the former equals the fixed service time in the latter. For which queue would you expect the mean queue length to be the shorter?

7.10. A queue is serviced by r servers, with the distribution of the inter-arrival times for the queue being exponential with parameter λ and the service times distributions for each server being exponential with parameter μ. If N is the random variable for the length of the queue *including* those being served, show that its expected value is

$$\mathbf{E}(N) = p_0\left[\sum_{n=1}^{r-1}\frac{\rho^n}{(n-1)!} + \frac{\rho^r[r^2 + \rho(1-r)]}{(r-1)!(r-\rho)^2}\right],$$

where $\rho = \lambda/\mu < 1$, and

$$p_0 = 1\left/\left[\sum_{n=0}^{r-1}\frac{\rho^n}{n!} + \frac{\rho^r}{(r-\rho)(r-1)!}\right]\right.$$

(see Eqn (7.11)).

If $r = 2$, show that

$$\mathbf{E}(N) = \frac{4\rho}{4-\rho^2}.$$

For what interval of values of ρ is the expected length of the queue less than the number of servers?

7.11. For a queue with two servers, the probability p_n that there are n servers in the queue, including those being served, is given by

$$p_0 = \frac{2-\rho}{2+\rho}, \qquad p_1 = \rho p_0, \qquad p_n = 2\left(\frac{\rho}{2}\right)^n p_0, \qquad (n \geq 2),$$

where $\rho = \lambda/\mu$ (see Section 7.4). If the random variable N is the number of people in the

---

[2] Pierre-Simon Laplace (1749–1827), French mathematician.

queue, find its probability generating function. Now find the mean length of the queue including those being served.

7.12. The queue $M(\lambda)/D(\tau)/1$, which has a fixed service time of duration τ for every customer, has the probability generating function

$$G(s) = \frac{(1 - \rho)(1 - s)}{1 - se^{\rho(1-s)}},$$

where $\rho = \lambda\tau$ $(0 < \rho < 1)$ (see Section 7.5).
(a) Find the probabilities p_0, p_1, p_2.
(b) Find the expected value and variance of the length of the queue.
(c) Customers are allowed a service time τ which is such that the expected length of the queue is two individuals. Find the value of the traffic density ρ.

7.13. For the baulked queue that has a maximum length of m beyond which customers are turned away, the probabilities that there are n individuals in the queue are given by

$$p_n = \frac{\rho^n(1 - \rho)}{1 - \rho^{m+1}}, \quad (0 < \rho < 1), \quad p_n = \frac{1}{m + 1} \quad (\rho = 1),$$

for $n = 0, 1, 2, \ldots, m$. Show that the probability generating functions are

$$G(s) = \frac{(1 - \rho)[1 - (\rho s)^{m+1}]}{(1 - \rho^{m+1})(1 - \rho s)}, \quad (0 < \rho < 1),$$

and

$$G(s) = \frac{1 - s^{m+1}}{(m + 1)(1 - s)}, \quad (\rho = 1).$$

Find the expected value of the queue length including the person being served.

7.14. In Section 7.3(ii), the expected length of the queue with parameters λ and μ, including the person being served, was shown to be $\rho/(1 - \rho)$. What is the expected length of the queue *excluding* the person being served?

7.15. An $M(\lambda)/M(\mu)/1$ queue is observed over a long period of time. Regular sampling indicates that the mean length of the queue including the person being served is 3, whilst the mean waiting time to completion of service by any customer arriving is 10 minutes. What is the mean service time?

7.16. A person arrives at an $M(\lambda)/M(\mu)/1$ queue. If there are two people in the queue (including customers being served) the customer goes away and does not return. If there are fewer than two queueing then the customer joins the queue. Find the expected waiting time for the customer to the start of service.

7.17. A customer waits for service in a bank in which there are four counters with customers at each counter but otherwise no one is queueing. If the service time distribution is exponential with parameter μ for each counter, for how long should the queueing customer have to wait?

7.18. A hospital has a waiting list for two operating theatres dedicated to a particular group of operations. Assuming that the queue is in a steady state, and that the waiting list and operating time can be viewed as an $M(\lambda)/M(\mu)/2$ queue, show that the expected value of the random variable N representing the length of the queue is given by

$$\mathbf{E}(N) = \frac{\rho^3}{4 - \rho^2}, \qquad \rho = \frac{\lambda}{\mu} < 2.$$

The average waiting list is very long at 100 individuals. Why will ρ be very close to 2? Put $\rho = 2 - \varepsilon$ where $\varepsilon > 0$ is small. Show that $\varepsilon \approx 0.02$. A third operating theatre is brought into use with the same operating parameter μ. What effect will this new theatre have on the waiting list eventually?

7.19. Consider the $M(\lambda)/M(\mu)/r$ queue which has r servers such that $\rho = \lambda/\mu < r$. Adapting the method for the single-server queue (Section 7.3 (iii)), explain why the average service time for $(n-r+1)$ customers to be served is $(n-r+1)/(\mu r)$ if $n \geq r$. There are n persons in the queue including those being served. What is it if $n < r$? If $n \geq r$, show that the expected value of the waiting time random variable T until service is

$$\mathbf{E}(T) = \sum_{n=r}^{\infty} \frac{n - r + 1}{r\mu} p_n.$$

What is the average waiting time if service is included?

7.20. Consider the $M(\lambda)/M(\mu)/r$ queue. Assuming that $\lambda < r\mu$, what is the probability in the long-term that at any instant there is no one queueing excluding those being served?

7.21. Access to a toll road is controlled by a plaza of r toll booths. Vehicles approaching the toll booths choose one at random: any toll booth is equally likely to be the one chosen irrespective of the number of cars queueing (perhaps an unrealistic situation). The payment time is assumed to be exponential with parameter μ, and vehicles are assumed to approach as Poisson with parameter λ. Show that, viewed as a steady-state process, the queue of vehicles at any toll booth is an $M(\lambda/r)/M(\mu)/1$ queue assuming $\lambda/(r\mu) < 1$. Find the expected number of vehicles queueing at any toll booth. How many cars would you expect to be queueing over all booths?

One booth is out of action, and vehicles distribute themselves randomly over the remaining booths. Assuming that $\lambda/[(r - 1)\mu] < 1$, how many extra vehicles can be expected to be queueing overall?

7.22. In an $M(\lambda)/M(\mu)/1$ queue, it is decided that the service parameter μ should be adjusted to make the mean length of the busy period 10 times the slack period, to allow the server some respite. What should μ be in terms of λ in the steady-state process?

7.23. In the baulked queue (see Example 7.1) not more than $m \geq 2$ people (including the person being served) are allowed to form a queue, the arrivals having a Poisson distribution with parameter λ. If there are m individuals in the queue, then any further arrivals are turned away. It is assumed that the service distribution is exponential with rate μ. If $\rho = \lambda/\mu \neq 1$, show that the expected length of the busy periods is given by $(1 - \rho^m)/(\mu - \lambda)$.

7.24. The $M(\lambda)/D(\tau)/1$ queue has a fixed service time τ, and from Section 7.5, its probability generating function is

$$G(s) = \frac{(1 - \lambda\tau)(1 - s)}{1 - se^{\lambda\tau(1-s)}}.$$

Show that the expected length of its busy periods is $\tau/(1 - \lambda\tau)$.

7.25. A certain process has the $(r + 1)$ states E_n, $(n = 0, 1, 2, \ldots r)$. The transition rates between state n and state $n + 1$ is $\lambda_n = (r - n)\lambda$, $(n = 0, 1, 2, \ldots, r - 1)$, and between n and $n - 1$ is $\mu_n = n\mu$, $(n = 1, 2, \ldots, r)$. These are the only possible transitions at each step in the process. (This could be interpreted as a 'capped' birth and death process in which the population size cannot exceed r.)

Find the differential-difference equation for the probabilities $p_n(t)$, $(n = 0, 1, 2, \ldots, r)$, that the process is in state n at time t. Consider the corresponding steady-state process in which $dp_n/dt \to 0$ and $p_n(t) \to p_n$ as $t \to \infty$. Show that

$$p_n = \left(\frac{\lambda}{\mu}\right)^n \binom{r}{n} \frac{\mu^r}{(\lambda + \mu)^r}, \quad (n = 0, 1, 2, \ldots, r).$$

7.26. In a baulked single-server queue, not more than 3 spaces are allowed. Write down the full time-dependent equations for the probabilities $p_0(t), p_1(t), p_2(t), p_3(t)$. Show that

$$\mathbf{p}(t) = \begin{bmatrix} p_0(t) & p_1(t) & p_2(t) & p_3(t) \end{bmatrix}$$

satisfies

$$\frac{d\mathbf{p}(t)}{dt} = A\mathbf{p}(t) \quad \text{where} \quad A = \begin{bmatrix} -\lambda & \mu & 0 & 0 \\ \lambda & -(\lambda + \mu) & \mu & 0 \\ 0 & \lambda & -(\lambda + \mu) & 0 \\ 0 & 0 & \lambda & -\mu \end{bmatrix}.$$

Whilst the general problem can be solved, to make the algebra manageable, let $\mu = 2\lambda$. Find the eigenvalues and eigenvectors of A, and write down the general solution for $\mathbf{p}(t)$. If the queue starts with no customers at time $t = 0$, find the solution for $\mathbf{p}(t)$.

CHAPTER 8

Reliability and Renewal

8.1 Introduction

The expected life until breakdown occurs of domestic equipment such as television sets, DVD recorders, washing machines, central heating boilers, etc., is important to the consumer. Failure means the temporary loss of the facility and expense. A television set, for example, has a large number of components, and the failure of just one of them may cause the failure of the set. Increasingly, individual components now have high levels of reliability, and the reduction of the likelihood of failure by the inclusion of alternative components or circuits is probably not worth the expense for a television which will suffer some deterioration with time anyway. Also, failure of such equipment is unlikely to be life-threatening.

In other contexts, reliability is extremely important for safety or security reasons, as in, for example, aircraft or defence installations where failure could be catastrophic. In such systems back-up circuits and components and fail-safe systems are necessary, and there could still be the remote possibility of both main and secondary failure. Generally, questions of cost, safety, performance, reliability, and complexity have to be resolved together with risk assessment.

In medical situations the interest lies in the survivability of the patient, that is, the expected survival time to death or relapse, under perhaps alternative medical treatments. Again, factors such as cost and quality of life can be important.

Reliability is determined by the probability of a device giving a specified performance for an intended period. For a system we might be interested in the probability that it is still operating at time t, the expected time to failure, and the failure rate. There are many external factors which could affect these functions, such as manufacturing defects (which often cause early failure), unexpected operating conditions, or operator error.

8.2 The reliability function

Let T be a nonnegative random variable which is the time to failure or **failure time** for a component, device, or system. Suppose that the **distribution function** of T is F so that

$$
\begin{aligned}
F(t) &= \mathbf{P}\{\text{component fails at or before time } t\}, \\
&= \mathbf{P}\{T \le t\}, \quad t \ge 0.
\end{aligned}
$$

The **density** of T is denoted by f, that is,

$$f(t) = \frac{dF(t)}{dt}. \tag{8.1}$$

It is more convenient for the consumer or customer to look at the **reliability** or **survivor function**, $R(t)$, rather than the failure time distribution. The function $R(t)$ is the probability that the device is still operating after time t. Thus,

$$R(t) = \mathbf{P}\{T > t\} = 1 - F(t). \tag{8.2}$$

From (8.1),

$$\begin{aligned}
f(t) = \frac{dF(t)}{dt} &= \lim_{\delta t \to 0} \frac{F(t + \delta t) - F(t)}{\delta t}, \\
&= \lim_{\delta t \to 0} \frac{\mathbf{P}\{T \le t + \delta t\} - \mathbf{P}\{T \le t\}}{\delta t}, \\
&= \lim_{\delta t \to 0} \frac{\mathbf{P}\{t < T \le t + \delta t\}}{\delta t}.
\end{aligned}$$

Hence for small δt, $f(t)\delta t$ is approximately the probability that failure occurs in the time interval $(t, t + \delta t)$.

The **failure rate function** or **hazard function** $r(t)$ is defined as that function for which $r(t)\delta t$ is the conditional probability that failure occurs in the interval $(t, t + \delta t)$ *given that the device was operating at time* t. As the limit of this conditional probability, the failure rate function is given by

$$r(t) = \lim_{\delta t \to 0} \frac{\mathbf{P}\{t < T \le t + \delta t | T > t\}}{\delta t}. \tag{8.3}$$

There are two important results relating the reliability and failure rate functions, which will be be proved in the following theorem.

Theorem 8.1

(i) The failure rate function

$$r(t) = \frac{f(t)}{R(t)}. \tag{8.4}$$

(ii) The reliability function

$$R(t) = \exp\left[-\int_0^t r(s)ds\right]. \tag{8.5}$$

Proof (i) Use the conditional probability definition (Section 1.3):

$$\mathbf{P}\{A|B\} = \mathbf{P}\{A \cap B\}/\mathbf{P}\{B\}.$$

Thus

$$\begin{aligned}
\mathbf{P}\{t < T \le t + \delta t | T > t\} &= \frac{\mathbf{P}\{(t < T \le t + \delta t) \cap (T > t)\}}{\mathbf{P}\{T > t\}} \\
&= \frac{\mathbf{P}\{t < T \le t + \delta t\}}{\mathbf{P}\{T > t\}}.
\end{aligned}$$

From (8.2) and (8.3),

$$r(t) = \lim_{\delta t \to 0} \frac{\mathbf{P}\{t < T \leq t + \delta t\}}{\delta t} \frac{1}{\mathbf{P}\{T > t\}} = \frac{f(t)}{R(t)}.$$

(ii) Using (8.1) and the last result,

$$r(t) = \frac{f(t)}{R(t)} = \frac{dF(t)}{dt} \frac{1}{R(t)} = -\frac{dR(t)}{dt} \frac{1}{R(t)} = -\frac{d}{dt}(\ln R(t)).$$

Integration of this result gives

$$\ln R(t) = -\int_0^t r(s)ds,$$

since $R(0) = 1$ (the device is assumed to be operating at start-up). Hence

$$R(t) = \exp\left[-\int_0^t r(s)ds\right].$$

Example 8.1 *Whilst in use, an office photocopier is observed to have a failure rate function* $r(t) = 2\lambda t$ *per hour where* $\lambda = 0.00028(hours)^{-2}$; *in other words,* $r(t)$ *is the probability that the photocopier fails within the next hour given that it was working at the beginning of the hour at time t. What is the probability that the photocopier is still operational after* 20 *hours? Also find the associated probability density function* $f(t)$ *for the time to failure.*

The failure rate function increases linearly with time. By Theorem 8.1(ii),

$$R(t) = \exp\left[-\int_0^t r(s)ds\right] = \exp\left[-\int_0^t 2\lambda s ds\right] = \exp(-\lambda t^2)$$

Thus the probability that the photocopier is still working after 20 hours is

$$R(20) = e^{-0.00028 \times (20)^2} = 0.89.$$

By Theorem 8.1(i), the associated probability density function is

$$f(t) = r(t)R(t) = 2\lambda t e^{-\lambda t^2} \quad t \geq 0,$$

which is a Weibull distribution (see Section 1.8) with parameters λ and 2.

8.3 Exponential distribution and reliability

For the exponential distribution (see Section 1.8), the probability density function is $f(t) = \lambda e^{-\lambda t}$ $(t \geq 0)$. Thus, for the exponential distribution,

$$R(t) = 1 - F(t) = 1 - \int_0^t f(s)ds = 1 + [e^{-\lambda s}]_0^t = e^{-\lambda t}, \quad (8.6)$$

and

$$r(t) = \frac{f(t)}{R(t)} = \frac{\lambda e^{-\lambda t}}{e^{-\lambda t}} = \lambda.$$

This constant failure rate implies that the probability of failure in a time interval δt given survival at time t is independent of the time at which this occurs. Thus the conditional survival probability for times t beyond, say, time s is

$$\mathbf{P}\{T > s+t | T > s\} \quad = \quad \frac{\mathbf{P}\{(T > s+t) \cap (T > s)\}}{\mathbf{P}\{T > s\}} = \frac{\mathbf{P}\{T > s+t\}}{\mathbf{P}\{T > s\}}$$

$$= \quad \frac{e^{-\lambda(s+t)}}{e^{-\lambda s}} = e^{-\lambda t},$$

which is the no-memory property of the exponential distribution. The implication is that such devices do not deteriorate with age, which means that they can be re-used with the same probability of subsequent failure at any stage in their lives.

8.4 Mean time to failure

In terms of the probability density function, the mean time to failure is

$$\mathbf{E}(T) = \int_0^\infty t f(t) dt. \tag{8.7}$$

The mean can be expressed in terms of the reliability function $R(t)$ as follows. Since, from Eqn (8.1) and Eqn (8.2), $f(t) = -dR/dt$, then

$$\mathbf{E}(T) = -\int_0^\infty t \frac{dR(t)}{dt} dt = -[tR(t)]_0^\infty + \int_0^\infty R(t) dt$$

after integrating by parts. Assuming that $tR(t) \to 0$ as $t \to \infty$ (this is certainly true for the exponential distribution, since $R(t) = e^{-\lambda t}$, which tends to zero faster than $t \to \infty$), then

$$\mathbf{E}(T) = \int_0^\infty R(t) dt. \tag{8.8}$$

If $R(t)$ decays as $1/t$ or slower, then we have the seeming paradox that a system can fail although the mean time is unbounded.

For the exponential distribution of the previous section, $R(t) = e^{-\lambda t}$ and

$$\mathbf{E}(T) = \int_0^\infty e^{-\lambda t} dt = 1/\lambda. \tag{8.9}$$

Example 8.2 *Find the mean time to failure for the photocopier of Example 8.1. What is the variance of the failure times?*

It was shown that reliability function was $R(t) = e^{-\lambda t^2}$. Hence the mean time to failure is

$$\mathbf{E}(T) = \int_0^\infty e^{-\lambda t^2} dt = \tfrac{1}{2}\sqrt{(\pi/\lambda)} = 52.96 \text{ hours},$$

using the standard formula for the integral (see the Appendix).

Now

$$\mathbf{E}(T^2) \quad = \quad \int_0^\infty t^2 f(t) dt = -\int_0^\infty t^2 \frac{dR(t)}{dt} dt$$

$$= -[t^2 R(t)]_0^\infty + \int_0^\infty 2tR(t)dt = \int_0^\infty 2te^{-\lambda t^2} dt$$

$$= 1/\lambda,$$

since $t^2 R(t) = t^2 e^{-\lambda t^2} \to 0$ as $t \to \infty$. Hence the variance is given by

$$\mathbf{V}(T) = \mathbf{E}(T^2) - [\mathbf{E}(T)]^2 = \frac{1}{\lambda} - \frac{\pi}{4\lambda} = \frac{4-\pi}{4\lambda} = 766.4 \text{ (hours)}^2.$$

8.5 Reliability of series and parallel systems

In many systems, components can be in series configurations as in Figure 8.1(a), or in parallel configurations as in Figure 8.1(b), or in complex series/parallel systems as in Figure 8.1(c), perhaps with bridges as shown in Figure 8.1(d).

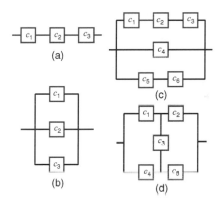

Figure 8.1 *(a) Components in series. (b) Components in parallel. (c) Mixed series/parallel system. (d) Mixed system with bridge.*

In the series of components in Figure 8.1(a), the system is only operable if all the components are working; if any one fails then the system fails. Suppose that the system has m components $c_1, c_2, \ldots, c_m$ in series with reliability functions $R_1(t)$, $R_2(t), \ldots, R_m(t)$. Whilst it is not generally true that components operate independently, since failure of one will affect the operation of others, we make the assumption of independence in deriving the results below. If T is the random variable of the time to failure in the system, and T_n is the random variable of the time to failure of component c_n, and assuming the independence of the T_n's, then

$$\begin{aligned} R(t) &= \mathbf{P}\{\text{ all } T > t\} \\ &= \mathbf{P}\{T_1 > t \text{ and } T_2 > t \text{ and} \cdots \text{ and } T_m > t\} \\ &= R_1(t)R_2(t)\ldots R_m(t) = \prod_{n=1}^{m} R_n(t), \end{aligned} \qquad (8.10)$$

which is simply the product of the individual reliability functions. The failure rate function

$$r(t) = -\frac{d}{dt} \ln R(t) = -\frac{d}{dt} \ln \prod_{n=1}^{m} R_n(t) = -\frac{d}{dt} \sum_{n=1}^{m} \ln R_n(t) = \sum_{n=1}^{m} r_n(t)$$
(8.11)

is the sum of the individual failure rate functions.

For the parallel configuration in Figure 8.1(b), it is easier to start from the distribution function of the time to failure $F(t)$. Since system failure can only occur when *all* components have failed,

$$
\begin{aligned}
F(t) &= \mathbf{P}\{T \le t\} = \mathbf{P}\{\text{system fails at or before time } t\}, \\
&= \mathbf{P}\{\text{all components fail at or before time } t\}, \\
&= \mathbf{P}\{T_1 \le t \text{ and } T_2 \le t \text{ and} \cdots \text{ and } T_m \le t\} \\
&= F_1(t)F_2(t)\dots F_m(t) = \prod_{n=1}^{m} F_n(t),
\end{aligned}
$$

where $F_n(t)$ is the distribution function of the time to failure of c_n. Consequently, the reliability function for m parallel components is

$$R(t) = 1 - F(t) = 1 - \prod_{n=1}^{m} F_n(t) = 1 - \prod_{n=1}^{m}[1 - R_n(t)].$$
(8.12)

A formula for $r(t)$ in terms of the individual functions $r_n(t)$ can be found but it is not particularly illuminating.

Example 8.3 *If each component c_n $(n = 1, 2, \dots, m)$ in a series arrangement has an exponentially distributed failure time with parameter λ_n, respectively, find the reliability function and the mean time to failure if the failure times are independent.*

The reliability function for c_n is $R_n(t) = e^{-\lambda_n t}$. Hence by Eqn (8.10), the reliability function for the series of components is

$$R(t) = \prod_{n=1}^{m} R_n(t) = e^{-\lambda_1 t} e^{-\lambda_2 t} \cdots e^{-\lambda_m t} = e^{-(\lambda_1 + \lambda_2 + \cdots + \lambda_m)t},$$

which is the same reliability function for that of a single component with exponentially distributed failure time with parameter $\sum_{n=1}^{m} \lambda_n$. Consequently by Eqn (8.9), the mean time to failure is $1/\sum_{n=1}^{m} \lambda_n$.

Example 8.4 *Three components c_1, c_2, c_3 in parallel have times to failure which are independent and identically exponential distributions with parameter λ. Find the mean time to failure.*

From (8.12) the reliability function is

$$R(t) = 1 - (1 - e^{-\lambda t})^3,$$

since $R_n(t) = e^{-\lambda t} (n = 1, 2, 3)$. The mean time to failure is

$$\mathbf{E}(T) = \int_0^{\infty} R(t)dt = \int_0^{\infty} \left[1 - (1 - e^{-\lambda t})^3\right] dt,$$

$$= \int_0^\infty \left[1 - (1 - 3e^{-\lambda t} + 3e^{-2\lambda t} - e^{-3\lambda t}) \right] dt,$$

$$= \int_0^\infty \left[3e^{-\lambda t} - 3e^{-2\lambda t} + e^{-3\lambda t} \right] dt$$

$$= \left[\frac{3}{\lambda} - \frac{3}{2\lambda} + \frac{1}{3\lambda} \right] = \frac{11}{6\lambda}.$$

For the mixed system in Figure 8.1(c), each series subsystem has its own reliability function. Thus the series c_1, c_2, c_3 has the reliability function $R_1(t)R_2(t)R_3(t)$ and c_5, c_6 the reliability function $R_5(t)R_6(t)$. These composite components are then in parallel with c_4. Hence, by (8.12), the reliability function for the system is

$$R(t) = 1 - (1 - R_1(t)R_2(t)R_3(t))(1 - R_4(t))(1 - R_5(t)R_6(t)).$$

To analyse the system in Figure 8.1(d) we return to a set approach. The easiest way to determine the reliability of this bridge system is to look at the bridge component c_3. Let S_i $(i = 1, 2, 3, 4, 5)$ be the event that c_i is operating.

If c_3 is working, then the system is active if either c_1 or c_4 or both are operating. Hence, in this case, the probability is

$$Q_1(t) - \mathbf{P}[(S_1 \cup S_4) \cap (S_2 \cup S_5)]. \tag{8.13}$$

If c_3 is not working, then the system is active if c_1 and c_2 are working, or c_4 and c_5 are working, or all are working: this probability is

$$Q_2(t) = \mathbf{P}[(S_1 \cap S_2) \cup (S_4 \cap S_5)]. \tag{8.14}$$

Hence the reliability function becomes

$$R(t) = Q_1(t)\mathbf{P}(S_3) + Q_2[1 - \mathbf{P}(S_3)]. \tag{8.15}$$

To simplify the algebra here suppose that the components c_1, c_2, c_4, and c_5 all have the *same* reliability function $R_c(t)$ whilst the bridge component has the reliability function $R_3(t)$. Hence, assuming independence and using the rules for union and intersection of sets,

$$\begin{aligned} Q_1(t) &= [\mathbf{P}(S_1) + \mathbf{P}(S_4) - \mathbf{P}(S_1)\mathbf{P}(S_4)][\mathbf{P}(S_2) + \mathbf{P}(S_5) - \mathbf{P}(S_2)\mathbf{P}(S_5)] \\ &= [2R_c(t) - R_c(t)^2]^2 \end{aligned}$$

$$\begin{aligned} Q_2(t) &= \mathbf{P}(S_1)\mathbf{P}(S_2) + \mathbf{P}(S_4)\mathbf{P}(S_5) - \mathbf{P}(S_1)\mathbf{P}(S_2)\mathbf{P}(S_4)\mathbf{P}(S_5) \\ &= 2R_c(t)^2 - R_c(t)^4. \end{aligned}$$

Finally, from (8.15), the reliability function for the bridge system is

$$R(t) = [2R_c(t) - R_c(t)^2]^2 R_3(t) + 2R_c(t)^2 - R_c(t)^4[1 - R_3(t)]. \tag{8.16}$$

The reliability function for the system with distinct components is given by the following result after much algebra:

$$R(t) = (1 - R_3(t))[R_1(t)R_2(t) + R_4(t)R_5(t) - R_1(t)R_2(t)R_4(t)R_5(t)]$$
$$+ R_3(t)[R_1(t) + R_4(t) - R_1(t)R_4(t)][R_2(t) + R_5(t) - R_2(t)R_5(t)].$$

For more details of the reliability of connected systems see Blake (1979, Ch. 12).

8.6 Renewal processes

Consider a system that has a single component which, when it fails, is replaced by an identical component with the same time to failure or **lifetime** probability distribution. It is assumed that there is no delay at replacement. In this **renewal process** we might ask: how many renewals do we expect in a given time t? Let $N(t)$ be the number of renewals up to and including time t (it is assumed that the first component is installed at start-up at time $t = 0$). Let T_r be a nonnegative random variable which represents the time to failure of the r-th component measured from its introduction. The time to failure of the first n components, S_n, is given by

$$S_n = T_1 + T_2 + \cdots + T_n.$$

At this point we include a diversion into joint distributions. We introduced joint probability distributions for discrete random variables in Section 1.10. A similar concept is required for continuous random variables. Consider the **random vector**

$$\mathbf{T} = (T_1, T_2, \ldots, T_n),$$

where $T_1, T_2, \ldots, T_n$ are each random variables. The **joint** or **multivariate distribution function** of $\mathbf{T}$ is defined by

$$F_T(t_1, t_2, \ldots, t_n) = \mathbf{P}(T_1 \leq t_1, T_2 \leq t_2, \ldots, T_n \leq t_n).$$

The individual or univariate **marginal distribution function** of T_i are

$$F(t_i) = \mathbf{P}(T_1 < \infty, T_2 < \infty, \ldots, T_i \leq t_i, \ldots, T_n < \infty).$$

The random variables are said to be **independent** if and only if

$$F_T(t_1, t_2, \ldots, t_n) = F_{T_1}(t_1) F_{T_2}(t_2) \cdots F_{T_n}(t_n).$$

The corresponding density function is

$$f_{T_1}(t_1) f_{T_2}(t_2) \cdots f_{T_n}(t_n),$$

where $f_{T_i}(t_i) = dF_{T_i}(t_i)/dt_i$, $(i, 1, 2, \ldots, n)$. However, in our discussion here we will only consider the special case in which all marginal distributions are identical with densities

$$f_{T_i}(t_i) = f(t_i),$$

and the T_i's are *iid*.

Consider first the case for $S_2 = T_1 + T_2$. For

$$F_2(s_2) = \mathbf{P}(T_1 + T_2 \leq s_2),$$

we must integrate the density over the (t_1, t_2)-plane such that $t_1 + t_2 \leq s_2$. Thus

$$\mathbf{P}\{T_1 + T_2 \leq s_2\} = \iint\limits_{\substack{x \geq 0, y \geq 0 \\ x + y \leq s_2}} f(x)f(y)dxdy,$$

$$= \int_0^{s_2} \int_0^{s_2 - y} f(x)f(y)dxdy,$$

since the joint density function is integrated over the triangle shown in Figure 8.2. Hence

$$\mathbf{P}\{T_1 + T_2 \leq s_2\} = \int_0^t F(s_2 - y)f(y)dy.$$

since

$$\int_0^{s_2 - y} f(y)dy = F(s_2 - y).$$

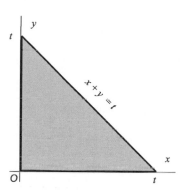

Figure 8.2 *Area of integration for nonnegative random variables.*

If we put

$$\mathbf{P}\{T_1 + T_2 \leq s_2\} = F_2(s_2),$$

then

$$F_2(s_2) = \int_0^{s_2} F_1(s_2 - y)f(y)dy, \qquad (8.17)$$

putting $F_1(t) = F(t)$ to suggest the iterative scheme. We can now repeat this process to obtain the next distribution,

$$F_3(s_3) = \mathbf{P}\{S_3 \leq s_3\} = \int_0^{s_3} F_2(s_3 - y)f(y)dy,$$

and, in general,

$$F_n(s_n) = \mathbf{P}\{S_n \leq s_n\} = \int_0^{s_n} F_{n-1}(s_n - y)f(y)dy. \qquad (8.18)$$

Example 8.5 *The lifetimes of components in a renewal process with instant renewal are independently and identically distributed with constant failure rate λ. Find the probability that at least two components have been replaced at time t.*

Since the components have constant failure rates, then they must have exponentially distributed failure times with parameter λ. Hence $f(t) = \lambda e^{-\lambda t}$ and $F(t) = 1 - e^{-\lambda t}$. Two components or more will have failed if $S_2 = T_1 + T_2 \leq t$. Hence by Eqn (8.17),

$$
\begin{aligned}
P\{S_2 \leq s_2\} = F_2(s_2) &= \int_0^{s_2} F(s_2 - y)f(y)dy = \int_0^{s_2} (1 - e^{-\lambda(s_2-y)})\lambda e^{-\lambda y}dy \\
&= \lambda \int_0^{s_2} (e^{-\lambda y} - e^{-\lambda s_2})dy, \\
&= \lambda \left[-\frac{e^{-\lambda y}}{\lambda} - ye^{-\lambda s_2} \right]_0^{s_2}, \\
&= 1 - e^{\lambda s_2} - \lambda s_2 e^{-\lambda s_2} = 1 - (1 + \lambda s_2)e^{-\lambda s_2}. \quad (8.19)
\end{aligned}
$$

The general formula for the density $f_n(t)$ is derived in the next section.

8.7 Expected number of renewals

We can find the probability that $N(t)$, the number of renewals up to and including t, is n, and the expected value of $N(t)$ as follows. The probability $\mathbf{P}\{N(t) = n\}$ can be expressed as the following difference:

$$
\mathbf{P}\{N(t) = n\} = \mathbf{P}\{N(t) \geq n\} - \mathbf{P}\{N(t) \geq n+1\}.
$$

We now use the important relation between S_n, the time to failure of the first n components, and $N(t)$, namely

$$
\mathbf{P}\{S_n \leq t\} = \mathbf{P}\{N(t) \geq n\}.
$$

This is true since if the total time to failure of n components is less than t, then there must have been at least n renewals up to time t. Hence,

$$
\mathbf{P}\{N(t) = n\} = \mathbf{P}\{S_n \leq t\} - \mathbf{P}\{S_{n+1} \leq t\} = F_n(t) - F_{n+1}(t),
$$

by (8.18). Thus the expected number of renewals by time t is

$$
\begin{aligned}
\mathbf{E}(N(t)) &= \sum_{n=1}^{\infty} n \left[F_n(t) - F_{n+1}(t) \right], \\
&= [F_1(t) + 2F_2(t) + 3F_3(t) + \cdots] - [F_2(t) + 2F_3(t) + \cdots], \\
&= F_1(t) + F_2(t) + F_3(t) + \cdots, \\
&= \sum_{n=1}^{\infty} F_n(t), \quad (8.20)
\end{aligned}
$$

which is simply the sum of the individual distribution functions which are given iteratively by Eqn (8.18). This is a general formula for any cumulative distribution function $F_n(t)$. The corresponding sequence of density functions are given by

$$
\begin{aligned}
f_n(t) &= \frac{dF_n(t)}{dt} = \frac{d}{dt} \int_0^t F_{n-1}(t - y)f(y)dy, \\
&= F_n(0)f(t) + \int_0^t F'_{n-1}(t - y)f(y)dy,
\end{aligned}
$$

$$= \int_0^t f_{n-1}(t-y)f(y)dy, \tag{8.21}$$

since $F_n(0) = 0$. Note that this is the same iterative formula as for the probability distributions given by Eqn (8.18).

Suppose now that the failure times are exponentially distributed with parameter λ: this is essentially a Poisson process. From Eqn (8.21),

$$f_2(t) = \frac{dF_2(t)}{dt} = \lambda^2 t e^{-\lambda t}, \quad t \geq 0. \tag{8.22}$$

Hence the density function $f_n(t)$ is a gamma distribution with parameters (n, λ). We can prove this by induction. Assume that

$$f_n(t) = \frac{\lambda^n}{(n-1)!} t^{n-1} e^{-\lambda t}.$$

Then, from (8.21),

$$\int_0^t \frac{\lambda^n}{(n-1)!} (t-y)^{n-1} e^{-\lambda(t-y)} \lambda e^{-\lambda y} dy = \frac{\lambda^{n+1}}{(n-1)!} e^{-\lambda t} \int_0^t (t-y)^{n-1} dy,$$

$$= \frac{\lambda^{n+1}}{n!} (t-y)^n e^{-\lambda t} = f_{n+1}(t).$$

Since the result has already been established for $f_2(t)$ in Eqn (8.22), the result is true by induction for $n = 3, 4, \ldots$. Thus,

$$\sum_{n=1}^{\infty} f_n(t) = \sum_{n=1}^{\infty} \frac{\lambda^n}{(n-1)!} t^{n-1} e^{-\lambda t} = \lambda e^{\lambda t} e^{-\lambda t} = \lambda. \tag{8.23}$$

Finally, from (8.23),

$$\mathbf{E}(N(t)) = \sum_{n=1}^{\infty} F_n(t) = \sum_{n=1}^{\infty} \int_0^t f_n(s) ds,$$

$$= \int_0^t \sum_{n=1}^{\infty} f_n(s) ds = \int_0^t \lambda ds = \lambda t,$$

a result which could have been anticipated since this is a Poisson process.

The expected value of $N(t)$ is known as the **renewal function** for the process.

8.8 Problems

8.1. The lifetime of a component has a uniform density function given by

$$f(t) = \begin{cases} 1/(t_1 - t_0) & 0 < t_0 < t < t_1 \\ 0 & \text{otherwise.} \end{cases}$$

For all $t > 0$, obtain the reliability function $R(t)$ and the failure rate function $r(t)$ for the component. Obtain the expected life of the component.

8.2. Find the reliability function $R(t)$ and the failure rate function $r(t)$ for the gamma density

$$f(t) = \lambda^2 t e^{-\lambda t}, \qquad t > 0.$$

How does $r(t)$ behave for large t? Find the mean and variance of the time to failure.

8.3. A failure rate function is given by

$$r(t) = \frac{t}{1 + t^2}, \qquad t \geq 0.$$

The rate of failures peaks at $t = 1$ and then declines towards zero as $t \to \infty$: failure becomes less likely with time (see Figure 8.3). Find the reliability function, and the corresponding probability density.

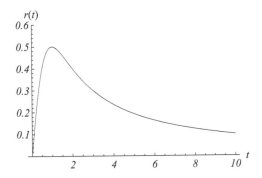

Figure 8.3 *Failure rate distribution $r(t)$.*

8.4. A piece of office equipment has a piecewise failure rate function given by

$$r(t) = \begin{cases} 2\lambda_1 t, & 0 < t \leq t_0, \\ 2(\lambda_1 - \lambda_2)t_0 + 2\lambda_2 t, & t > t_0 \end{cases} \qquad \lambda_1, \lambda_2 > 0.$$

Find its reliability function.

8.5. A laser printer is observed to have a failure rate function $r(t) = 2\lambda t$ per hour $(t > 0)$ whilst in use, where $\lambda = 0.00021 (\text{hours})^{-2}$: $r(t)$ is a measure of the probability of the printer failing in any hour given that it was operational at the beginning of the hour. What is the probability that the printer is working after 40 hours of use? Find the probability density function for the time to failure. What is the expected time before the printer will need maintenance?

8.6. The time to failure is assumed to be gamma with parameters α and n, that is, with

$$f(t) = \frac{\alpha(\alpha t)^{n-1} e^{-\alpha t}}{(n-1)!}, \qquad t > 0.$$

Show that the reliability function is given by

$$R(t) = e^{-\alpha t} \sum_{r=0}^{n-1} \frac{\alpha^r t^r}{r!}.$$

Find the failure rate function and show that $\lim_{t\to\infty} r(t) = \alpha$. What is the expected time to failure?

8.7. An electrical generator has an exponentially distributed failure time with parameter λ_f and the subsequent repair time is exponentially distributed with parameter λ_r, the times being independent. The generator is started up at time $t = 0$. What is the mean time for the generator to fail and the mean time from $t = 0$ for it to be operational again?

8.8. A hospital takes a grid supply of electricity which has a constant failure rate λ. This supply is backed up by a stand-by generator which has a gamma distributed failure time with parameters $(2, \mu)$. Find the reliability function $R(t)$ for the whole electricity supply. Assuming that time is measured in hours, what should the relation between the parameters λ and μ be in order that $R(1000) = 0.999$?

8.9. The components in a renewal process with instant renewal are identical with constant failure rate $\lambda = (1/50)(\text{hours})^{-1}$. If the system has one spare component which can take over when the first fails, find the probability that the system is operational for at least 24 hours. How many spares should be carried to ensure that continuous operation for 24 hours occurs with probability 0.98?

8.10. A device contains two components c_1 and c_2 with independent failure times T_1 and T_2 from time $t = 0$. If the densities of the times to failure are f_1 and f_2 with distribution functions F_1 and F_2, show that the probability that c_1 fails before c_2 is given by

$$P\{T_1 < T_2\} = \int_{y=0}^{\infty}\int_{x=0}^{y} f_1(x)f_2(y)dxdy,$$

$$= \int_{y=0}^{\infty} F_1(y)f_2(y)dy.$$

Find the probability $P\{T_1 < T_2\}$ in the cases:
(a) both failure times are exponentially distributed with parameters λ_1 and λ_2;
(b) both failure times have gamma distributions with parameters $(2, \lambda_1)$ and $(2, \lambda_2)$.

8.11. Let T be the failure time of a component. Suppose that the density

$$f(t) = \alpha_1 e^{-\lambda_1 t} + \alpha_2 e^{-\lambda_2 t}, \qquad \alpha_1, \alpha_2 > 0, \qquad \lambda_1, \lambda_2 > 0,$$

where the parameters satisfy

$$\frac{\alpha_1}{\lambda_1} + \frac{\alpha_2}{\lambda_2} = 1.$$

Find the reliability function $R(t)$ and the failure rate function $r(t)$ for this 'double' exponential distribution. How does $r(t)$ behave as $t \to \infty$?

8.12. The lifetimes of components in a renewal process with instant renewal are *iid* with constant failure rate λ. Find the probability that at least three components have been replaced by time t.

8.13. The lifetimes of components in a renewal process with instant renewal are independent and identically distributed, and independent with a failure rate which has a uniform distribution with density

$$f(t) = \begin{cases} 1/k & 0 < t < k \\ 0 & \text{elsewhere.} \end{cases}$$

Find the probability that at least two components have been replaced at time t.

8.14. The lifetimes of components in a renewal process with instant renewal are identically distributed and independent, each with reliability function

$$R(t) = \tfrac{1}{2}(e^{-\lambda t} + e^{-2\lambda t}), \qquad t \geq 0, \qquad \lambda > 0.$$

Find the probability that at least two components have been replaced by time t.

8.15. The random variable T is the time to failure from $t = 0$ of a system. The distribution function for T is $F(t)$, $(t > 0)$. Suppose that the system is still functioning at time $t = t_0$. Let T_{t_0} be the conditional time to failure from this time, and let $F_{t_0}(t)$ be its distribution function. Show that

$$F_{t_0}(t) = \frac{F(t + t_0) - F(t_0)}{1 - F(t_0)},$$

and that the mean of T_{t_0} is

$$\mathbf{E}(T_{t_0}) = \frac{1}{1 - F(t_0)} \int_{t_0}^{\infty} [1 - F(u)]\,du.$$

8.16. Suppose that the time to failure, T of a system is uniformly distributed. Using the result from Problem 8.15, find the conditional distribution function assuming that the system is still working at time $t = t_0$.

8.17. In the bridge system represented by Figure 8.1(d), suppose that *all* components have the same reliability function $R_c(t)$. Show that the reliability function $R(t)$ is given by

$$R(t) = 2R_c(t)^2 + 2R_c(t)^3 - 5R_c(t)^4 + 2R_c(t)^5.$$

Suppose that the bridge c_3 is removed, What is the reliability function $R_x(t)$ for this system? Show that

$$R(t) > R_x(t).$$

What does this inequality imply?

CHAPTER 9

Branching and Other Random Processes

9.1 Introduction

In this chapter we look at other stochastic processes including branching and epidemic models. Branching processes are concerned with the generational growth and decay of populations. The populations could be mutant genes, neutrons in a nuclear chain reaction, or birds or animals which have annual cycles of births. As the name implies a branching process creates a **tree** with branches which can split into other branches at each step or at each generation in a chain. In this chapter we shall look at a simplified problem in which the process starts from a single individual which generates the tree. Such a process models an application such as cellular growth rather than an animal population in which both births and deaths are taking place continuously in time.

9.2 Generational growth

Consider a model in which a single individual (cell, organism) has known probabilities of producing a given number of descendants at a given time, and produces no further descendants. In turn these descendants each produce further descendants at the next subsequent time with the same probabilities. The process carries on in the same way, creating successive **generations**, as indicated in Figure 9.1. At each step there is a probability p_j that any individual creates j descendants $(j = 0, 1, 2, \ldots)$,

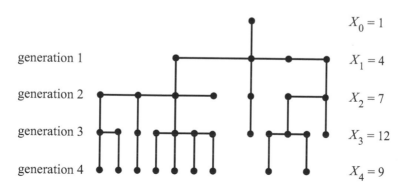

Figure 9.1 *Generational growth in a branching process.*

183

and it is assumed that this probability distribution $\{p_j\}$ is the same for every individual at every generation. We are interested in the population size in generation n: the earlier generations have either died out, or an individual can be viewed as a continuation of itself, or are not counted in the process.

Let X_n be a random variable representing the population size of generation n. Since the process starts with one individual, then $X_0 = 1$. In the illustration of a particular process in Figure 9.1,

$$X_1 = 3, \qquad X_2 = 7, \qquad X_3 = 12, \qquad \ldots$$

As described, the generations evolve in step, and occur at the same time steps. Note that populations do not necessarily increase with the generations.

Suppose that the probability distribution of descendant numbers $\{p_j\}$ has the probability generating function

$$G(s) = \sum_{j=0}^{\infty} p_j s^j. \tag{9.1}$$

($G(s)$ will be $G_1(s)$ later, but the suffix is often suppressed.) Since $X_0 = 1$, $G(s)$ is the generating function for the random variable X_1. For the second generation it is assumed that each descendant has the same probability p_j of independently creating j descendants. Let $G_2(s)$ be the generating function of the random variable X_2, which is the sum of X_1 random variables (the number of descendants of X_0), which, in turn, are denoted by the independent random variables $Y_1, Y_2, \ldots, Y_{X_1}$ (assumed to be identically distributed), so that

$$X_2 = Y_1 + Y_2 + \cdots + Y_{X_1}.$$

It might help understanding to interpret the random variables as given in Fig 9.1. For example, $X_2 = Y_1 + Y_2 + Y_3 + Y_4$, and the actual readings for the Y's are, from the left, $y_1 = 4$, $y_2 = 1$, $y_3 = 0$, $y_4 = 2$.

Returning to the general exposition,

$$P(Y_k = j) = p_j, \qquad P(X_1 = m) = p_m.$$

Let

$$P(X_2 = n) = h_n.$$

Then, using the partition law (the law of total probability) (see Section 1.3),

$$\begin{aligned} h_n &= P(X_2 = n) = \sum_{r=0}^{\infty} P(X_2 = n | X_1 = r) P(X_1 = r) \\ &= \sum_{r=0}^{\infty} p_r P(X_2 = n | X_1 = r). \end{aligned}$$

Now multiply h_n by s^n and sum over n to create the generating function called $G_2(s)$:

$$G_2(s) = \sum_{n=0}^{\infty} h_n s^n = \sum_{r=0}^{\infty} p_r \sum_{n=0}^{\infty} P(X_2 = n | X_1 = r) s^n. \tag{9.2}$$

Consider the latter summation in which r is a fixed integer. Then, by Section 1.9, where the probability generating function is defined,

$$\sum_{n=0}^{\infty} P(X_2 = n | X_1 = r) s^n \quad = \quad \mathbf{E}(s^{X_2} | X_1 = r) = \mathbf{E}(s^{Y_1 + Y_2 + \cdots + Y_r})$$

$$= \quad \mathbf{E}(s^{Y_1}) \mathbf{E}(s^{Y_2}) \dots \mathbf{E}(s^{Y_r}) \quad \text{(by independence)}$$

$$= \quad G(s) G(s) \dots G(s) \quad \text{(multiplied } r \text{ times)}$$

$$= \quad [G(s)]^r,$$

since $\mathbf{E}(s^{Y_1}) = G(s)$, $\mathbf{E}(s^{Y_2}) = G(s)$, and so on (see Section 1.9). This result also follows from the observation that the distribution of X_2 is the **convolution** of the distribution $\{p_j\}$ with itself r times. Finally, Eqn (9.2) becomes

$$G_2(s) = \sum_{r=0}^{\infty} p_r [G(s)]^r = G(G(s)). \tag{9.3}$$

Since $G(1) = 1$, it follows that $G_2(1) = G(G(1)) = G(1) = 1$.

This result holds between any two successive generations. Thus if $G_m(s)$ is the probability generating function of X_m, then

$$G_m(s) = G_{m-1}(G(s)) = G(G(\dots (G(s)) \dots)), \tag{9.4}$$

which has $G(s)$ nested m times on the right-hand side. This general result will not be proved here. In this type of branching process, $G(s)$ is the probability generating function of the probability distribution of the numbers of descendants from any individual in any generation to the next generation. It follows that $G_m(1) = 1$ for $m = 3, 4, 5, \dots$ by extending the result above.

Example 9.1. *Suppose in a branching process that any individual has a probability $p_j = 1/2^{j+1}$ $(j = 0, 1, 2, \dots)$ of producing j descendants in the next generation. Find the generating function for X_n, the random variable representing the number of descendants in the n-th generation given that $X_0 = 1$.*

From (9.1), the probability generating function for X_1 is

$$G(s) = \sum_{j=0}^{\infty} p_j s^j = \sum_{j=0}^{\infty} \frac{1}{2^{j+1}} s^j = \frac{1}{2 - s},$$

after summing the geometric series. From (9.3),

$$G_2(s) = \frac{1}{2 - G(s)} = \frac{1}{2 - \frac{1}{2-s}} = \frac{1}{2 - \frac{1}{2-s}} = \frac{2 - s}{3 - 2s},$$

$$G_3(s) = \frac{2 - G(s)}{3 - 2G(s)} = \frac{3 - 2s}{4 - 3s}, \qquad G_4(s) = \frac{4 - 3s}{5 - 4s}.$$

Generally

$$G_n(s) = G(G_{n-1}(s)) = \frac{1}{2 - G_{n-1}(s)} = \frac{1}{2 - \frac{1}{2 - G_{n-2}(s)}} = \cdots,$$

which is an example of a **continued fraction**.

Looking at the expressions for $G_2(s)$, $G_3(s)$, and $G_4(s)$, we might speculate that

$$G_n(s) = \frac{n - (n-1)s}{n+1-ns}.$$

This result can be established by an induction argument outlined as follows. Assume that the formula holds for n. Then

$$
\begin{aligned}
G(G_n(s)) &= \frac{1}{2 - [n - (n-1)s]/[n+1-ns]} \\
&= \frac{n+1-ns}{2[n+1-ns] - [n - (n-1)s]} \\
&= \frac{n+1-ns}{n+2-(n+1)s},
\end{aligned}
$$

which is the formula for $G_{n+1}(s)$, that is, n is replaced by $n+1$ in the expression for $G_n(s)$. Hence, by induction on the integers, the result is true since we have confirmed it directly for $n = 2$ and $n = 3$.

The power series expansion for $G_n(s)$ can be found by using the binomial theorem. Thus

$$
\begin{aligned}
G_n(s) &= \frac{n - (n-1)s}{n+1-ns} = \frac{n-(n-1)s}{n+1}\left[1 - \frac{ns}{n+1}\right]^{-1} \\
&= \frac{1}{n+1}[n - (n-1)s]\sum_{r=0}^{\infty}\left(\frac{ns}{n+1}\right)^r \\
&= \sum_{r=0}^{\infty}\left(\frac{n}{n+1}\right)^{r+1}s^r - \sum_{r=0}^{\infty}\left(\frac{n-1}{n+1}\right)\left(\frac{n}{n+1}\right)^r s^{r+1} \\
&= \frac{n}{n+1} + \sum_{r=1}^{\infty}\frac{n^{r-1}}{(n+1)^{r+1}}s^r.
\end{aligned}
$$

The probability that the population of generation n has size r is the coefficient of s^r, in this series, which is

$$\frac{n^{r-1}}{(n+1)^{r+1}} \quad \text{for} \quad r \geq 1.$$

The probability of extinction for generation n is $G_n(0) = n/(n+1)$, and this approaches 1 as $n \to \infty$, which means that the probability of ultimate extinction is certain for this branching process.

9.3 Mean and variance

The mean and variance of the population size of the n-th generation can be obtained as fairly simple general formulae as functions of the mean and variance of X_1, the random variable of the population size of the first generation.

Let μ_n and σ_n^2 be the mean and variance of the size of the n-th generation. As before let $G(s)$ be the generating function of the first generation. Then (as in Section 1.9),

$$\mu_1 = \mathbf{E}(X_1) = G'(1), \tag{9.5}$$

$$\sigma_1^2 = \mathbf{V}(X_1) = G''(1) + G'(1) - G'(1)^2 = G''(1) + \mu_1 - \mu_1^2. \tag{9.6}$$

From the previous section, the probability generating function of X_2 is $G(G(s))$. Then

$$\mu_2 = \frac{d}{ds}G_2(s)|_{s=1} = \frac{d}{ds}[G(G(s))]_{s=1}$$

$$= \left[\frac{d}{dG}G(G) \cdot \frac{d}{ds}G(s)\right]_{s=1}$$

$$= G'(1)G'(1) = \mu_1^2,$$

using the chain rule in calculus.

The method can be repeated for $E(X_3)$, $E(X_4)$, ... Thus, μ_n, the mean population size of the n-th generation, is given by

$$\mu_n = E(X_n) = \frac{dG_n(s)}{ds}\bigg|_{s=1} = \frac{d[G(G_{n-1}(s))]}{ds}\bigg|_{s=1}$$

$$= G'(1)G'_{n-1}(1) = \mu_1\mu_{n-1} = \mu_1^2\mu_{n-2} = \cdots = \mu_1^n \quad (9.7)$$

since $G_{n-1}(1) = 1$.

The variance of the population size of the n-th generation is σ_n^2, say, where

$$\sigma_n^2 = V(X_n) = G_n''(1) + G_n'(1) - [G_n'(1)]^2 = G_n''(1) + \mu_1^n - \mu_1^{2n}. \quad (9.8)$$

We can obtain a formula for σ_n^2 as follows. Differentiate

$$G_n'(s) = G'(G_{n-1}(s))G_{n-1}'(s)$$

again with respect to s:

$$G_n''(s) = G''(G_{n-1}(s))[G_{n-1}'(s)]^2 + G'(G_{n-1}(s))G_{n-1}''(s),$$

so that, with $s = 1$,

$$G_n''(1) = G''(G_{n-1}(1))[G_{n-1}'(1)]^2 + G'(G_{n-1}(1))G_{n-1}''(1)$$

$$= G''(1)[G_{n-1}'(1)]^2 + G'(1)G_{n-1}''(1)$$

$$= (\sigma_1^2 - \mu_1 + \mu_1^2)\mu_1^{2n-2} + \mu_1 G_{n-1}''(1), \quad (9.9)$$

using Section 1.9(c). Equation (9.9) is a first-order inhomogeneous linear difference equation for $G_n''(1)$.

Write the equation as

$$G_n''(1) - \mu_1 G_{n-1}''(1) = (\sigma_1^2 - \mu_1 + \mu_1^2)\mu_1^{2n-2}.$$

There are two cases to consider separately: $\mu_1 \neq 1$ and $\mu_1 = 1$.

(i) $\mu_1 \neq 1$. The corresponding homogeneous equation to (9.9)

$$G_n''(1) - \mu_1 G_{n-1}''(1) = 0$$

has the solution $G_n''(1) = B\mu_1^n$, where B is a constant. For the particular solution find the constant C such that $G_n''(1) = C\mu_1^{2n}$ satisfies the equation. It is easy to show that

$$C = \frac{\sigma_1^2 - \mu_1 + \mu_1^2}{\mu_1(\mu_1 - 1)}.$$

Hence

$$G_n''(1) = B\mu_1^n + \frac{(\sigma_1^2 - \mu_1 + \mu_1^2)\mu_1^{2n}}{\mu_1(\mu_1 - 1)}.$$

Since for the first generation ($n = 1$), $G''(1) = \sigma_1^2 - \mu_1 + \mu_1^2$, it follows that

$$B = -\frac{\sigma_1^2 - \mu_1 + \mu_1^2}{\mu_1(\mu_1 - 1)},$$

so that

$$G_n''(1) = \frac{\mu_1^n(\sigma_1^2 - \mu_1 + \mu_1^2)(\mu_1^n - 1)}{\mu_1(\mu_1 - 1)}, \qquad (9.10)$$

where it is now obvious why μ_1 cannot be 1 in this formula. Finally from Eqns (9.8) and (9.10),

$$\sigma_n^2 = G_n''(1) + \mu_1^n - \mu_1^{2n} = \frac{\sigma_1^2 \mu_1^{n-1}(\mu_1^n - 1)}{\mu_1 - 1}.$$

(ii) $\mu_1 = 1$. The equation becomes

$$G_n''(1) - G_{n-1}''(1) = \sigma_1^2.$$

The general solution of this difference equation is now

$$G_n''(1) = A + \sigma_1^2 n.$$

Since $G''(1) = \sigma_1^2$, it follows that $A = 0$. Hence, from Eqn (9.8),

$$\sigma_n^2 = G_n''(1) = n\sigma_1^2. \qquad (9.11)$$

Example 9.2. *For the branching process with probability $p_j = 1/2^{j+1}$, $j = 0, 1, 2, \ldots$ that any individual has j descendants in the next generation (see Example 9.1), find the mean and variance of the population size of the n-th generation.*

From Example 9.1, the probability generating function of X_1 is $G(s) = 1/(2 - s)$. Its mean is $\mu_1 = G'(1) = 1$. Hence by (9.7), the mean value of the size of the population of the n-th generation is $\mu_n = \mu_1^n = 1$.

The variance of X_1 is given by (see Eqn (9.5))

$$\sigma_1^2 = G''(1) + \mu_1 - \mu_1^2 = 2.$$

Hence, by (9.11), $\sigma_n^2 = 2n$.

Note that, although the probability of ultimate extinction is certain for this branching process, the mean remains 1 for all n. The variance shows increasing spread of values as n increases. This result shows that a fixed mean and an increasing variance or dispersal increases the probability of extinction.

9.4 Probability of extinction

In Example 9.1, the probability of ultimate extinction for a branching process with the probabilities $p_j = 1/2^{j+1}$ was shown to be certain for this particular process. A general result concerning possible extinction of a population can be found from the probability generating function $G(s)$ for X_1, the random variable of the population size of the first generation.

The probability of extinction by generation n is $G_n(0)$, where, as a reminder, $G_n(s)$ is the generating function of the n-th generation. Let $g_n = G_n(0)$, $(n = 1, 2, , \ldots))$. Thus g_1 is the probability of extinction in the first generation. Since

$$G_1(s)(= G(s)) = \sum_{n=0}^{\infty} p_n s^n,$$

then $g_1 = G_1(0) = p_0$.

We can make some general observations about the function $G(s)$. Since $0 \leq s \leq 1$ and $0 \leq p_n \leq 1$ for all n,

(i) $G(s) > 0$ for $p_0 > 0$;

(ii) $G'(s) = \sum_{n=1}^{\infty} n p_n s^{n-1} \geq 0$;

(iii) $G''(s) = \sum_{n=2}^{\infty} n(n-1) p_n s^{n-2} \geq 0$,

assuming that the series in (ii) and (iii) converge. Results (ii) and (iii) imply that the positive function $G(s)$ has an increasing slope with increasing s in $0 \leq s \leq 1$. These conditions define what is known as a **convex function** as illustrated in Figure 9.2. The graph of a convex function has the property that the curve (for $0 \leq s \leq 1$)

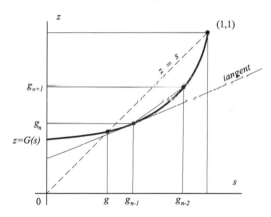

Figure 9.2 *The convex function $z = G(s)$ drawn in the case $\mu_1 > 1$.*

always *lies above every tangent* to the curve.

The sequence of extinction probabilities $\{g_n\}$ satisfies

$$g_n = G_n(0) = G(G_{n-1}(0)) = G(g_{n-1}), \qquad (n = 2, 3, \ldots). \qquad (9.12)$$

In particular,

$$g_2 = G_2(0) = G(G(0)) = G(g_1) = \sum_{n=0}^{\infty} p_n s^n > p_0 = G(0) = g_1. \qquad (9.13)$$

As stated above, the convexity of $G(s)$ implies that the slope of the **chord** joining

$(g_{n-1}, G(g_{n-1}))$ and $(g_n, G(g_n))$ must be greater than the slope of the tangent at $(g_{n-1}, G(g_{n-1}))$, that is,

$$\frac{G(g_n) - G(g_{n-1})}{g_n - g_{n-1}} > G'(g_{n-1}), \qquad \text{or} \qquad \frac{g_{n+1} - g_n}{g_n - g_{n-1}} > 0$$

by (9.12) and (9.13). Therefore if $g_n > g_{n-1}$, then $g_{n+1} > g_n$. From (9.13), $g_2 > g_1$: hence, by induction on the integers, $\{g_n\}$ is an increasing sequence.

This increasing sequence is *bounded* above since $g_n \leq 1$ for all n. We now appeal to a theorem in mathematical analysis which (briefly) states that a *bounded increasing sequence tends to a limit*, say g in this case. Hence

$$g_n \to g \quad \text{as} \quad n \to \infty.$$

Since $g_n = G(g_{n-1})$, the limits of both sides imply that g satisfies

$$g = G(g). \tag{9.14}$$

In Figure 9.2 the solutions of (9.14) can be represented by intersections of the line $z = s$ and the curve $z = G(s)$. The two always intersect at $(1, 1)$ but there may be a further intersection in the interval $(0 < s < 1)$. The critical value of the slope $G'(s)$ which divides one solution from two occurs where the line $z = s$ is tangential to $z = G(s)$ at $s = 1$: in other words if $G'(1) = 1$. Therefore

if $G'(1) > 1$ then $g_n \to g < 1$;

if $G'(1) \leq 1$ then $g_n \to 1$.

However, $G'(1)$ is the mean μ_1 of the first generation. Hence we have the following early simple check on extinction:

if $\mu_1 > 1$, then ultimate extinction occurs with probability g;

if $\mu_1 \leq 1$, then ultimate extinction is certain.

Example 9.3. *For a certain branching process with $X_0 = 1$, the probability generating function for X_1 is given by*

$$G(s) = \frac{1}{(2 - s)^2}.$$

Find the probability that the population ultimately becomes extinct. Also find the mean and variance of the populations in the n-th generation. Interpret the results.

The probability of ultimate extinction is the smallest positive root of

$$g = G(g) = \frac{1}{(2 - g)^2}.$$

Graphs of $z = s$ and $z = 1/(2 - s)^2$ are shown in Figure 9.3, which has one point of intersection for $0 < s < 1$, and the expected intersection at $(1, 1)$. The solution for g satisfies

$$g(2 - g)^2 = 1, \quad \text{or} \quad (g - 1)(g^2 - 3g + 1) = 0.$$

The required solution is

$$g = \frac{1}{2}(3 - \sqrt{5}) \approx 0.382,$$

which is the probability of ultimate extinction

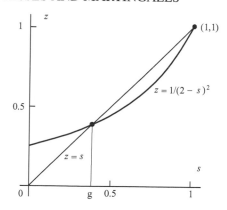

Figure 9.3 *Graphs of $z = s$ and $z = 1/(2 - s)^2$.*

Since

$$G'(s) = \frac{2}{2 - s)^3}, \qquad G''(s) = \frac{6}{(2 - s)^4},$$

then

$$\mu_1 = G'(1) = 2, \qquad \sigma_1^2 = G''(1) + \mu_1 - \mu_1^2 = 6 + 2 - 4 = 4.$$

Hence by (9.6) and (9.11), the mean and variance of the n-th generation are given by

$$\mu_n = \mu_1^n = 2^n, \qquad \sigma_n^2 = \frac{\sigma_1^2 \mu_1^{n-1}(\mu_1^n - 1)}{\mu_1 - 1} = 2^{n+1}(2^n - 1).$$

The mean, which behaves as 2^n, tends to infinity much faster than the variance, which behaves as n. The spread is not increasing as fast as the mean. This implies that, as n increases, the population either becomes extinct with probability of approximately 0.382 or becomes very large with probability approximately $1 - 0.382 = 0.618$: the probability of modest population sizes is increasingly very small.

9.5 Branching processes and martingales

Consider again the branching process starting with one individual so that $X_0 = 1$, and with $G(s)$ as the probability generating function of X_1. As before, X_n is the random variable representing the population size of the n-th generation. Suppose we look at the expected value of the random variable X_{n+1} given the random variables $X_0, X_1, X_2, \ldots, X_n$. By Section 1.10, such a conditional expectation is not a number but itself a *random variable*. The number of descendants at step n is the random variable X_n, each of which will have a mean number of descendants $\mu_1 = G'(1)$, since the probability of creating descendants is the same at each generation. Hence the expectation of X_{n+1} given dependent on the previous random variables is the product of X_n and the mean, that is

$$\mathbf{E}(X_{n+1}|X_0, X_1, X_2, \ldots, X_n) = X_n \mu_1. \tag{9.15}$$

Let

$$Z_n = \frac{X_n}{\mathbf{E}(X_n)} = \frac{X_n}{\mu_1^n}$$

using (9.7). The random variable Z_n is the random variable X_n *normalised* by its own mean. Hence, using (9.7) again

$$\mathbf{E}(Z_n) = \mathbf{E}\left(\frac{X_n}{\mu_1^n}\right) = \frac{\mathbf{E}(X_n)}{\mu_1^n} = 1 = \mathbf{E}(X_0),$$

since the process starts with one individual in this case. Hence (9.15) becomes

$$\mathbf{E}(Z_{n+1}\mu_1^{n+1}|X_0, X_1, X_2, \ldots, X_n) = Z_n\mu_1^{n+1},$$

or

$$\mathbf{E}(Z_{n+1}|X_0, X_1, X_2, \ldots, X_n) = Z_n,$$

since $\mathbf{E}(aX) = a\mathbf{E}(X)$ (see Section 1.6), where a is a constant and X is any random variable. Such a random variable sequence $\{Z_n\}$, in which the expected value of Z_{n+1} conditional on random variable sequence $\{X_n\}$ is Z_n, is known as a **martingale**[1]. In some problems as in the case discussed below the conditioning sequence can be the *same* sequence of random variables, and may be thought of as a **self-conditioning martingale**.

The most famous martingale arises in the following gambling problem. A gambler makes an even money bet with a casino starting with a pound bet. If she or he wins, the gambler has her or his £1 stake returned plus £1 from the casino. If the gambler loses the pound to the casino, she or he then bets £2, £4, £8,..., until she or he wins. Suppose the gambler first wins at the n-th bet. Then the gambler will have won £2^n for an outlay of

$$£(1 + 2 + 2^2 + 2^3 + \cdots + 2^{n-1}) = £(2^n - 1),$$

summing the geometric series. Hence the gambler always wins £1 at some stage. It is a guaranteed method of winning but does require a large financial backing, and for obvious reasons casinos do not always permit this form of gambling. The martingale betting scheme will always beat what is a fair game in terms of the odds at each play.

We shall now explain why this gamble is a martingale. Suppose that the gambler starts with £1 and bets against the casino according to the rules outlined above. The doubling bets continue irrespective of the outcome at each bet. Let Z_n be the gambler's total asset or debt at the n-th bet: $Z_0 = 1$ and Z_n can be a negative number indicating that the gambler owes money to the casino. Since $Z_0 = 1$, then Z_1 is a random variable which can take the values 0 or 2. Given Z_0 and Z_1, Z_2 can take any one of the values $-2, 0, 2, 4$, while Z_3 can take any one of the values $-6, -4, -2, 0, 2, 4, 6, 8$, and so on as shown by the **tree** in Figure 9.4. The random variable Z_n can take any one of the values

$$\{-2^n + 2, -2^n + 4, \ldots, 2^n - 2, 2^n\}, \qquad (n = 1, 2, \ldots),$$

---

[1] The word has an obscure French origin. Apart from the doubling stake gamble in eighteenth century France, it also means a strap in a horse's harness, and a rope in ship's rigging.

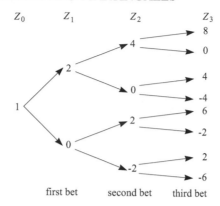

Figure 9.4 *Possible outcomes of the first three bets in the gambling martingale.*

or, equivalently, in increasing order,

$$Z_n = \{-2^n + 2m + 2\} \qquad (m = 0, 1, 2, \ldots, 2^n - 1)$$

for $n = 1, 2, 3, \ldots$. The probability of any one of these values occurring at the n-th bet is $1/2^n$. It follows that the expected value of Z_n (the gambler's mean asset) is

$$\mathbf{E}(Z_n) = \sum_{m=0}^{2^n-1} \frac{1}{2^n}(-2^n + 2m + 2) = \frac{1}{2^n}(-2^n + 2)\sum_{m=0}^{2^n-1} 1 + \frac{2}{2^n}\sum_{m=1}^{2^n-1} m = 1,$$

using the arithmetic sum

$$\sum_{m=1}^{2^n-1} m = \frac{1}{2}2^n(2^n - 1).$$

The expected winnings in a game of doubling bets would be £1 irrespective of the duration of the wager. Tactically it would be best to play until the first win, and then start a new game with £1.

The conditional expectation of Z_3 given Z_0, Z_1, and Z_2 (remember these conditional expectations are random variables) is the random variable with outcomes

$$\mathbf{E}(Z_3|Z_0, Z_1, Z_2) = \{-2, 0, 2, 4\} = Z_2,$$

as shown in Figure 9.4. The mean values of the pairs in the final column give the corresponding numbers in the previous column. Generally

$$\mathbf{E}(Z_{n+1}|Z_0, Z_1, Z_2, \ldots, Z_n) = Z_n,$$

which is a martingale *with respect to its own random variable sequence.*

Example 9.4 *Let the random variable X_n be the position of a walker in a symmetric random walk after n steps. Suppose that the walk starts at the origin so that $X_0 = 0$. Show that $\{X_n\}$ is a martingale with respect to its own sequence of random variables. (See Section 3.2 for discussion of the properties of the symmetric random walk.)*

In terms of the modified Bernoulli random variable,

$$X_{n+1} = X_0 + \sum_{i=1}^{n+1} W_i = X_n + W_{n+1},$$

since $X_0 = 0$. By the Markov property of the random walk,

$$\mathbf{E}(X_{n+1}|X_0, X_1, \ldots, X_n) = \mathbf{E}(X_{n+1}|X_n).$$

Suppose that $X_n = k$ where k must satisfy $-n \le k \le n$. The random variable W_{n+1} can take either of the values -1 or $+1$, each with probability of $\frac{1}{2}$, so that $X_{n+1} = k - 1$ if $W_{n+1} = -1$ or $X_{n+1} = k + 1$ if $W_{n+1} = +1$. Since these outcomes are equally likely, it follows that

$$\mathbf{E}(X_{n+1}|X_n = k) = \frac{1}{2}(k - 1) + \frac{1}{2}(k + 1) = k = X_n,$$

which defines a martingale with respect to itself.

Example 9.5. *Let $X_1, X_2, \ldots$ be independent identically distributed random variables, each of which can take the values -1 and $+1$ with probabilities of $\frac{1}{2}$. Show that the partial sums*

$$Z_n = \sum_{j=1}^{n} \frac{1}{j} X_j, \quad (n = 1, 2, \ldots),$$

form a martingale with respect to $\{X_n\}$.

It follows that

$$
\begin{aligned}
\mathbf{E}(Z_{n+1}|X_1, X_2, \ldots, X_n) &= \mathbf{E}\left(Z_n + \frac{1}{n+1} X_{n+1} | X_1, \ldots, X_n\right) \\
&= \mathbf{E}(Z_n|X_1, \ldots, X_n) + \frac{1}{n+1} \mathbf{E}(X_{n+1}|X_1, \ldots, X_n) \\
&= Z_n + \frac{1}{n+1} \mathbf{E}(X_{n+1}) = Z_n,
\end{aligned}
$$

since $\mathbf{E}(X_{n+1}) = 0$.

The random variable Z_n is the partial sum of a harmonic series with terms $1/j$ with randomly selected signs, known as a random harmonic series. It is known that the harmonic series

$$\sum_{j=1}^{\infty} \frac{1}{j} = 1 + \frac{1}{2} + \frac{1}{3} + \frac{1}{4} + \cdots$$

diverges and that the alternating harmonic series

$$\sum_{j=1}^{\infty} (-1)^{j+1} \frac{1}{j} = 1 - \frac{1}{2} + \frac{1}{3} - \frac{1}{4} + \cdots$$

converges. It can be shown that the random harmonic series converges with probability 1. This requires considerable extra theory: a full account can be found in Lawler (1995).

9.6 Stopping rules

Suppose we have a random process, say a Markov chain or a branching process, with random variables $\{X_r\}$, $(r = 0, 1, 2, \ldots)$ in which we decide to stop the process when a specified condition is met. Hence $\{X_r\}$ may go through a sequence of values $(x_0, x_1, \ldots, x_n)$ and then stop at x_n when a predetermined stopping condition is met. The random variable T for the **stopping time** or number of steps can take any of the integer values $(0, 1, 2, \ldots)$. A simple example of a stopping rule is that $n = m$ where m is a given positive integer: in other words the process stops after m steps. In the case just given, T, of course, is known, but X_m is not, in which case the expected value $\mathbf{E}(X_m)$ might be of interest. Alternatively, we might wish to find the expected value $\mathbf{E}(T)$ for a problem in which the stopping rule is that the process stops when $X_n \geq x$ where x is a given number.

We could say that if X_m is a 'reward', then $\mathbf{E}(X_m)$ could be used to determine at which point to stop playing.

Example 9.6. *A random walk starts at the origin, with probability p that the walk advances one step and probability $q = 1 - p$ that the walk retreats one step at every position. The walk stops after $T = 10$ steps. What is the expected position of the walk at the stopping time?*

In terms of the modified Bernoulli random variable $\{W_i\}$ (see Section 3.2), the random variable of the position of the walk after n steps is

$$X_n = X_0 + \sum_{i=1}^{n} W_i = \sum_{i=1}^{n} W_i$$

since $X_0 = 0$. It was shown in Section 3.2 that

$$\mathbf{E}(X_n) = n(p - q) = n(2p - 1).$$

Hence for $T = 10$, the expected position is

$$\mathbf{E}(X_{10}) = 10(p - q).$$

Another stopping rule could be that the process stops when the random variable X_n first takes a particular value or one of several values. This could be a particular strategy in a gambling game where the gambler abandons the game when certain winnings are achieved or cuts his or her losses when a certain deficit is reached.

Example 9.7. *A symmetric random walk starts at position k. Let X_n $(n = 0, 1, 2, \ldots)$ be the random variable of the position of the walk at step n. Show that the random variable $Y_n = X_n^2 - n$ is a martingale with respect to X_n.*

This slightly surprising result needs an explanation. The table of possible outcomes is:

$X_0 = k$;

$X_1 = k + 1, k - 1$;

$X_2 = k + 2, k, k, k - 2$;

$X_3 = k + 3, k + 1, k + 1, k - 1, k + 1, k - 1, k - 1, k - 3$;

$\cdots \qquad \cdots$

That there are two values of k for X_2 occurs because k can be reached by different routes, and similarly for repeats in succeeding random variables. The list of outcomes for X_n^2 is:

$$X_0^2 = k^2;$$
$$X_1^2 = (k+1)^2, (k-1)^2;$$
$$X_2^2 = (k+2)^2, k^2, k^2, (k-2)^2;$$
$$X_3^2 = (k+3)^2, (k+1)^2, (k+1)^2, (k-1)^2, (k+1)^2, (k-1)^2, (k-1)^2, (k-3)^2;$$
$$\cdots \qquad \cdots$$

It is evident that $\{X_n^2|X_{n-1}\}$ is not a martingale since, for example, $\mathbf{E}(X_2^2|X_1 = k)$ has the possible outcomes

$$\{\tfrac{1}{2}[(k+2)^2 + k^2], \tfrac{1}{2}[k^2 + (k-2)^2]\} = \{(k+1)^2 + 1, (k-1)^2 + 1\},$$

which is not X_1^2 (see list above). However,

$$\begin{aligned}
\mathbf{E}(Y_2|X_1) &= \mathbf{E}(X_2^2 - 1|X_1) = \{\tfrac{1}{2}[(k+2)^2 + k^2 - 2], \tfrac{1}{2}[k^2 + (k-1)^2 - 2]\} \\
&= \{(k+1)^2, (k-1)^2\} \\
&= Y_1.
\end{aligned}$$

Generally,

$$\mathbf{E}(Y_n|X_{n-1}) = Y_{n-1}, \qquad (n = 3, 4, \ldots),$$

which defines a martingale. This result provides an alternative derivation of the duration of the game in the gambler's ruin (see Chapter 2).

We showed in Example 9.4 that $\{X_n\}$ is a martingale with respect to the modified Bernoulli random variable $\{W_n\}$ for the symmetric random walk. It has just been shown that

$$\mathbf{E}(X_T) = k = X_0 = \mathbf{E}(X_0),$$

since X_0 is specified in this problem. For this martingale and for certain other martingales generally, this result is not accidental. In fact, the **stopping theorem** states:

If $\{Z_n\}$ is a martingale with respect to $\{X_n\}$ ($n = 0, 1, 2, \ldots$), T is a defined stopping time, and
(i) $\mathbf{P}(T < \infty) = 1$,
(ii) $\mathbf{E}(|Z_T|) < \infty$,
(iii) $\mathbf{E}(Z_n|T > n)\mathbf{P}(T > n) \to 0$ as $n \to \infty$,
then $\mathbf{E}(Z_T) = \mathbf{E}(Z_0)$.

This is a useful result, which we shall not prove here (see Grimmett and Stirzaker (1982), p. 209, or, Lawler (2006), p. 93).

Example 9.8. *As in the previous example a symmetric random walk starts at k. Obtain the duration of the walk which terminates when either 0 or a is first reached This is equivalent to the gambler's ruin problem in Chapter 2.*

The stopping time T occurs when either 0 or a is first reached. We intend to find $\mathbf{E}(T)$, for which we shall use the random variable $Y_T = X_T^2 - n$ where X_n and Y_n were defined in the previous example. From the stopping theorem,

$$\mathbf{E}(Y_T) = \mathbf{E}(Y_0) = \mathbf{E}(X_0^2 - 0) = k^2, \qquad (9.16)$$

from the first list in Example 9.7. Also,

$$
\begin{aligned}
\mathbf{E}(Y_T) &= \mathbf{E}(X_T^2 - T) = \mathbf{E}(X_T^2) - \mathbf{E}(T) \\
&= 0 \times \mathbf{P}(X_T = 0) + a^2 \times \mathbf{P}(X_T = a) - \mathbf{E}(T) \\
&= a^2 \times \frac{k}{a} - \mathbf{E}(T) = ak - \mathbf{E}(T),
\end{aligned}
\tag{9.17}
$$

where $\mathbf{P}(X_T = a) = v_{a-k} = k/a$ (see Section 2.2 for the derivation of v_a). Therefore, from (9.16) and (9.17),

$$
\mathbf{E}(T) = ak - \mathbf{E}(Y_T) = ak - k^2 = k(a - k),
$$

which is an alternative derivation of the expected duration d_k of Section 2.4.

Example 9.9 *(Polya's urn[2]). An urn contains a red ball and a green ball. A ball is removed at random from the urn and the ball is returned to the urn together with a further ball of the same colour. This procedure is repeated. After the first replacement the urn contains the three balls, and after the n-th replacement $n + 2$ balls. Let X_n be the random variable of the number of green balls at this stage. Show the random variable $M_n = X_n/(n + 2)$ is a martingale.*

At step n the urn contains x_n (say) green balls and $n + 2 - x_n$ red balls. The probability of choosing a green ball from the urn is $x_n/(n+2)$ and of a red ball $(n+2-x_n)/(n+2)$. Note that the random variables X_n form a Markov chain. After the n-th step there are either $X_n + 1$ or X_n green balls, that is, either a green or red ball is chosen, respectively. Suppose that we do not know *a priori* the form of the martingale but suspect it is of the form $M_n = u(n)X_n$ where $u(n)$ is to be determined. Hence, the conditional probability is

$$
\begin{aligned}
\mathbf{E}(M_{n+1}|X_n) &= u(n+1)(X_n + 1)\frac{X_n}{n+2} + u(n+1)X_n \frac{n+2-X_n}{n+2} \\
&= \frac{u(n+1)}{n+2}X_n + u(n+1)X_n \\
&= \frac{(n+3)u(n+1)}{n+2}X_n = M_n
\end{aligned}
$$

if $u(n) = 1/(n + 2)$, which means that M_n defines a martingale.

9.7 A continuous time epidemic

The subject of epidemiology from a theoretical viewpoint is concerned with the construction of stochastic models which can represent the spread of a specific disease through a population. It is a topic of particular current interest with the apparent advent of new virulent diseases which in some cases have no curative treatment. Some diseases are cyclical, such as measles and influenza; some diseases are currently spreading, such as HIV infections, which have long periods in which infected persons can transmit the disease to others; some diseases have disappeared, such as smallpox, and some which have declined through treatment (such as tuberculosis) are recurring through drug-resistant strains. The main aim of probability models is to be able to predict the likely extent of a disease—how many are infected through the period of the epidemic and how might inoculation affect the spread of the disease.

---

[2] George Polya (1887–1985) Hungarian mathematician.

There is also interest in the geographical spread, demographics, and time behaviour of diseases.

Modeling of epidemics is important in forming animal and public health policies, for example in outbreaks of foot and mouth disease, and the studies of the possible transmission of tuberculosis from badgers to cattle. These require more complicated models than we are able to discuss in this book. An indication of the complexity of such models may be found in Ferguson *et al.* (2001) and Cox *et al.* (2005). In the next section we shall develop a stochastic epidemic model which, however, requires numerical computation.

In this section we shall develop a model for the **simple epidemic**. In epidemiology there are assumed to be three groups of individuals in a population. These are the **susceptibles**, those who might succumb to the disease, the **infectives**, those who have the disease and can spread it among the susceptibles, and what we might lump together as the **immunes**, which includes, in addition to those who are immune, the isolated, the dead (as a result of the disease), and the recovered and now immune.

In practice the spread of diseases is more complicated than simple stochastic models display. They can be affected by many factors. Here is a fairly general description of the progress of a typical disease. If someone becomes infected there is usually a **latent period** during which the disease develops. At the end of this period the patient might display symptoms and become an **infective** although the two events might not occur at the same time. The individual remains an infective during the **infectious period** after which death or recovery occurs, and perhaps the individual becomes immune to further bouts of the illness.

In this simple epidemic the assumptions are very restrictive. It is assumed that the population contains only susceptibles and infectives—individuals do not die or recover but remain infective. This model is not very realistic but it might cover the early stages of a disease which has a long infectious period. Suppose that the population remains fixed at $n_0 + 1$ with no births or deaths during the epidemic and that one infective is introduced at time $t = 0$. Initially there are n_0 susceptibles. Let the random variable $S(t)$ represent the number of susceptibles which are *not* infected at time t. At this time there will be $n_0 + 1 - S(t)$ infectives. It is assumed that individuals in the population mix homogeneously, which is rarely the case except possibly with animals under controlled laboratory conditions. In the simple epidemic the main assumption is that the likelihood of an infection occurring in a time interval δt is

$$\beta S(t)[n_0 + 1 - S(t)]\delta t,$$

that is, it is proportional to the product of the numbers of susceptibles and infectives at time t. In this joint dependence the likelihood of infection will be high if both populations are relatively large, and small if they are both small. The constant β is known as the **contact rate**. The probability of more than one infection taking place in time δt is assumed to be negligible. Let $p_n(t)$, $0 \leq n \leq n_0$ be the probability that there are n susceptibles at time t. Then by the partition law,

$$p_n(t + \delta t) = \beta(n + 1)(n_0 - n)p_{n+1}(t)\delta t + [1 - \beta n(n_0 + 1 - n)\delta t]p_n(t) + o(\delta t),$$

for $0 \le n \le n_0 - 1$, and

$$p_{n_0}(t + \delta t) = (1 - \beta n_0 \delta t)p_{n_0}(t) + o(\delta t)$$

for $n = n_0$. In the first equation either one susceptible became an infective in time δt with probability $\beta(n + 1)(n_0 - n)\delta t$ when the population was $n + 1$ at time t, or no infection occurred with probability $1 - \beta n(n_0 + 1 - n)\delta t$ when the population was n. Following the usual method of dividing through by δt and then letting $\delta t \to 0$, we arrive at the differential-difference equations for the simple epidemic:

$$\frac{dp_n(t)}{dt} = \beta(n + 1)(n_0 - n)p_{n+1}(t) - \beta n(n_0 + 1 - n)p_n(t), \qquad (9.18)$$

for $n = 0, 1, 2, \ldots n_0 - 1$ and

$$\frac{dp_{n_0}(t)}{dt} = -\beta n_0 p_{n_0}(t). \qquad (9.19)$$

Since there are n_0 susceptibles at time $t = 0$, the initial condition must be $p_{n_0}(0) = 1$ with $p_n(0) = 0$ for $n = 0, 1, 2, \ldots, n_0 - 1$.

The partial differential equation for the probability generating function for these differential-difference equation is given in Problem 9.26, but unfortunately the equation does not seem to have a direct solution (see Bailey (1964), p. 173).

9.8 A discrete time epidemic model

In this section we shall introduce a **discrete time Markov chain** to model a simple epidemic. As in the previous, this epidemic has two populations, represented by the random variable N_S for the susceptibes and by N_I for the infectives. It is assumed that the total population remains fixed so that $N_S + N_I = n$ where n is a constant (we can assume that any births or deaths from other causes cancel one another, and there is no immigration). We shall also compare the results with a corresponding **deterministic model** which will be explained later.

We shall assume that any population changes of N_S and N_I can only take place at the discrete time steps of t at $\{0, \tau, 2\tau, 3\tau, \ldots\}$. Both N_S and N_I can take the values

$$N_S, N_I \in (0, 1, 2, \ldots, n).$$

We shall only consider N_I since $N_S = n - N_I$ in this model.

This process will be represented by a Markov chain over the time steps

$$\{0, \tau, 2\tau, 3\tau, \ldots\}.$$

We will switch to the Markov chain notation of Chapter 4 and let E_r ($r = 0, 1, 2, \ldots n$ be the event that the infective population is r. The transition diagram is as shown in Figure 9.5. In a Markov process only transitions between neighbouring states can occur: infections of two or more individuals or loss of infection of two or more in any time interval are assumed to be negligible. Hence the following transitions are possible:

- $E_i \to E_{i-1}, E_i, E_{i+1}$ ($i = 1, 2, \ldots, n - 1$).
- $E_n \to E_n, E_{n-1}$.

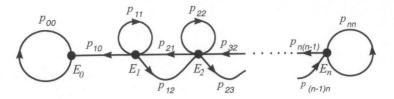

Figure 9.5 *A discrete-time Markov chain for an epidemic model.*

- $E_0 \to E_0$ (an absorbing state).

We now define the probabilities associated with the transitions denoted by p_{ij}, where

$$
\left.
\begin{array}{ll}
p_{i(i-1)} = i\gamma\tau, & (i = 1, 2, \dots, n) \\
p_{ii} = 1 - [\{\beta i(n - i)\tau/n\} + \gamma i\tau], & (i = 0, 1, 2, \dots, n) \\
p_{i(i+1)} = \beta i(n - i)\tau/n, & (i = 0, 1, 2, \dots, n - 1) \\
p_{ij} = 0, & \text{elsewhere}
\end{array}
\right\}
\tag{9.20}
$$

where γ and β are contact rates. The probability of a new infection, for example, is $p_{i(i+1)} = \beta i(n - i)\tau/n$, which depends on the joint numbers of infectives and susceptibles. The probability of recovery is $p_{i(i-1)} = \gamma i\tau$, which is proportional to the period τ of the time step. The other probabilities can be interpreted similarly. Further requirements are that $0 \le p_{ij} \le 1$, and the usual

$$
\sum_{j=0}^{n} p_{ij} = 1
$$

which is the case here. Finally the increment τ should be chosen so that the *maximum* value of p_{ij} for *all* i, j, β, γ, and n should be small and certainly less than 1. The transition probabilities p_{ij} are functions of the increments τ but are independent of time.

The transition matrix defined by (9.20) is a **tridiagonal matrix**. It is important to visualise how the matrix appears. This 5×5 version for $n = 4$ (population size) is the 5×5 matrix

$$
T = [p_{ij}] =
\begin{bmatrix}
1 & 0 & 0 & 0 & 0 \\
\gamma\tau & 1 - (\frac{3}{4}\beta + \gamma)\tau & \frac{3}{4}\beta\tau & 0 & 0 \\
0 & 2\gamma\tau & 1 - (\beta + 2\gamma)\tau & \beta\tau & 0 \\
0 & 0 & 3\gamma\tau & 1 - (\frac{3}{4}\beta + 3\gamma)\tau & \frac{3}{4}\beta\tau \\
0 & 0 & 0 & 4\gamma\tau & 1 - 4\gamma\tau
\end{bmatrix}.
$$

Returning to the general case we consider an example with $n = 100$, the total population size, and assume that there are 2 infectives[3] at time $t = 0$, meaning that the epidemic starts in E_2. Expressed as the vector this is

$$
\mathbf{s}_0 = \begin{bmatrix} 0 & 0 & 1 & 0 & \cdots 0 \end{bmatrix}.
$$

---

[3] Two infectives are chosen to avoid the epidemic dying out quickly: however, other initial values can be chosen if required.

At time $t = \tau$ the infective population will be in states E_1, E_2, or E_3 with probabilities given by

$$\mathbf{s}_0 T = \begin{bmatrix} 0 & 2\gamma\tau & 1 - [\{2\beta(n-2)/n\} + 2\gamma]\tau & 2\{\beta(n-2)/n\} & \cdots & 0 \end{bmatrix}.$$

Hence the probability that the infective population falls to 1 is $2\gamma\tau$, increases to 3 is $2\{\beta(n-2)/n\}$, or remains unchanged at 2 is the remaining probability.

If the transition is to E_3, then we set

$$\mathbf{s}_1 = \begin{bmatrix} 0 & 0 & 0 & 1 & \cdots 0 \end{bmatrix}.$$

If the transition is either to E_1 or remains in E_2, then we set the $\mathbf{s}_1$ accordingly. The next step is to compute $\mathbf{s}_1 T$ to determine the next state or event. The process is repeated so that we generate a sequence of infective population sizes together with the probabilities with which they occur.

If the final state E_n is reached (unlikely with the input data defined later) the whole population would be infected. However, since individuals can become uninfected and reinfected in this simple model, the process can continue indefinitely.

An infected population sequence as shown in Figure 9.6 have been computed for $E_2 = 1$ initially (2 infectives), and the parameters are $n = 100$, $\beta = 1$, $\gamma = 0.5$, $\tau = 0.01$. A check of the probabilities gives

$$\max([\beta j(n-j)\tau]/n) = \tfrac{1}{4}\beta n\tau = 0.25, \quad \max(\gamma j\tau) = 0.25,$$

which is comfortably less than 1. A computed output[4] for an infected population over 2,000 time steps is shown in Figure 9.6. It can be seen from the outputs that ultimately

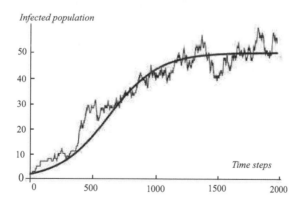

Figure 9.6 *A simulated stochastic output for an epidemic with* $n = 100$, $\beta = 1$, $\gamma = 0.5$, *time step* $\tau = 0.01$, *with initially 2 infectives. The smooth curve is the deterministic solution given by (9.24) with the same parameters and initial value.*

the populations seem to be hovering at about 50. It can be seen from (9.20) why this should occur: the up and down probabilities balance if

$$\beta j(n-j)\tau/n = \gamma j\tau \text{ or } j = (\beta - \gamma)n/\beta = 50,$$

---

[4] A program using *Mathematica* is available on the associated website.

in the example here. We shall say more about this in the next section. Note also that there is a finite, although extremely small, probability that E_0 is reached, in which case the epidemic ends. For an extensive account of this and other stochastic epidemic models, see Allen (2008).

9.9 Deterministic epidemic models

In the deterministic approach it is assumed that the susceptible and infective populations are continuous smooth functions of the time t controlled by **differential equations** (see Bailey (1964) and Allen (2008)). Let the continuous-time susceptible and infective populations be denoted by $x_S(t)$ and $x_I(t)$. The deterministic model is

$$\frac{dx_S}{dt} = -\frac{\beta}{n}x_Sx_I + \gamma x_I, \qquad (9.21)$$

$$\frac{dx_I}{dt} = \frac{\beta}{n}x_Sx_I - \gamma x_I, \qquad (9.22)$$

in which the argument t is dropped for simplicity. The assumption in the model is that the susceptible population in (9.21) decreases according to the joint contact between the susceptibles and the infectives and increases through individuals recovering at rate γ. The change in infectives is the mirror image of this. We have deliberately chosen the same symbols as in the stochastic case for the contact rates β and γ to emphasise the comparison. Addition of (9.21) and (9.22),

$$\frac{d(x_S + x_I)}{dt} = 0,$$

implies that $x_S + x_I = n$, that is, the total populations remains a constant n. Elimination of x_S in (9.22) leads to

$$\frac{dx_I}{dt} = x_I[(\beta - \gamma)n - \beta x_I]/n. \qquad (9.23)$$

This is a first-order differential equation which can be integrated directly to give, using partial fractions, with C a constant,

$$
\begin{aligned}
t + C &= \int \frac{n dx_I}{x_I[(\beta - \gamma)n - \beta x_I]} \\
&= \frac{1}{(\beta - \gamma)} \left[\int \frac{dx_I}{x_I} + \int \frac{\beta dx_I}{(\beta - \gamma)n - \beta x_I} \right] \\
&= \frac{1}{(\beta - \gamma)} \ln \left[\frac{x_I}{(\beta - \gamma)n - \beta x_I} \right],
\end{aligned}
$$

where we have assumed that $(\beta - \gamma)n > \beta x_I$ for all n and x_I. The solution of this equation for x_I is therefore

$$x_I = \frac{(\beta - \gamma)n}{\beta + e^{-(t+C)(\beta - \gamma)}}.$$

As in the stochastic model, we assume the initial condition $x_I(0) = 2$ so that

$$e^{-Cn(\beta - \gamma)} = \tfrac{1}{2}n(\beta - \gamma) - \beta,$$

and, finally

$$x_I = \frac{2(\beta - \gamma)n}{2\beta + [(\beta - \gamma)n - 2\beta]e^{-(\beta-\gamma)t}}. \tag{9.24}$$

As $t \to \infty$,

$$x_I \to \frac{(\beta - \gamma)n}{\beta}$$

($\beta > \gamma$ from the previous inequality). This limit is interesting since using the parameters $\beta = 1$, $\gamma = 0.5$ and population $n = 100$, as in the stochastic model in the previous section, then

$$x_I \to \tfrac{1}{2}n = 50,$$

if $n = 100$, agrees with the stochastic behaviour shown in Figure 9.6. The deterministic solution is also shown in the figure (in comparing the stochastic data) and the deterministic solution t must be replaced by $\tau t = 0.01t$ since the abscissa measures *time steps*.

Usually epidemics are recorded by the number of new cases which occur in set intervals (weeks, months, etc). From (9.24) this is given by

$$\frac{dx_I}{dt} = \frac{2n(\beta - \gamma)^2[(\beta - \gamma)n - 2\beta]e^{-(\beta-\gamma)t}}{[2\beta + \{(\beta - \gamma)n - 2\beta\}e^{-(\beta-\gamma)t}]^2}$$

in this case (of 2 infectives initially). The graph of the **epidemic curve**, as it is known, is shown in Figure 9.7.

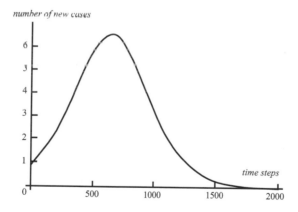

Figure 9.7 *Epidemic curve of number of new cases, dx_I/dt, against time steps with the same data: $\beta = 1$; $\gamma = 0.5$; $n = 100$. The epidemic peaks after about 700 time steps at a infective number of between six and seven individuals.*

There are other more elaborate models. In one of these immunity is included so that there are individuals who have contracted the disease but who subsequently become immune or are removed. This model includes this third population denoted by $x_R(t)$. The differential equations are (see Allen (2008)):

$$\frac{dx_S}{dt} = -\frac{\beta}{n}x_S Z_I + \nu(x_I + x_R), \tag{9.25}$$

$$\frac{dx_I}{dt} = \frac{\beta}{n}x_S x_I - \gamma x_I, \tag{9.26}$$

$$\frac{dx_R}{dt} = (\gamma - \nu)x_I - \nu x_R, \tag{9.27}$$

where β, ν, and γ are constant contact rates. This is still a model with a constant population

$$x_S + x_I + x_R = n. \tag{9.28}$$

It is assumed that births and deaths cancel. To interpret Eqn (9.25), for example, the rate of decrease of susceptibles behaves jointly with the infectives and susceptibles, but increases with recovered infectives and the immune. Also, recovered infectives can become susceptible again, and there are no deaths as a result of the epidemic.

There is an equilibrium state when all the derivatives are zero, given by the solutions of

$$-\frac{\beta}{n}x_S x_I + \nu(x_I + x_R) = 0,$$

$$\frac{\beta}{n}x_S x_I - \gamma x_I = 0,$$

$$(\gamma - \nu)x_I - \nu x_R = 0,$$

together with (9.28). The equilibrium state of the epidemic is given by

$$x_S = \frac{\gamma n}{\beta}, \quad x_I = \frac{n\nu(\beta - \gamma)}{\beta\gamma}, \quad x_R = \frac{n(\gamma - \nu)(\beta - \gamma)}{\beta\gamma},$$

achieved as $t \to \infty$, provided $\beta > \gamma > \nu$. If $\beta < \gamma$, then $Z_I \to 0$ as $t \to \infty$ and the infection dies out.

The deterministic model (9.25)–(9.27) does not have a simple solution, but it is possible to solve the equation numerically, and also to construct a stochastic model including immunity (see Allen (2008)).

9.10 An iterative solution scheme for the simple epidemic

As was remarked previously (Section 9.7), a partial differential equation for the probability generating function for Eqns (9.18) and (9.19) can be obtained, but the equation has no simple solution for the epidemic problem. Here we show how solutions for $p_n(t)$ can be constructed by an iterative procedure, although they do become increasingly complicated and they do not suggest an obvious general formula. It is possible to obtain explicit formulae for epidemics in very small populations.

Eliminate the parameter β by letting $\tau = \beta t$, by putting

$$p_n(\tau/\beta) = q_n(\tau) \tag{9.29}$$

so that Eqns (9.18) and (9.19) become the differential-difference equations

$$\frac{dq_n(\tau)}{d\tau} = (n+1)(n_0 - n)q_{n+1}(\tau) - n(n_0 + 1 - n)q_n(\tau), \quad (0 \le n \le n_0 - 1), \tag{9.30}$$

$$\frac{dq_{n_0}(\tau)}{d\tau} = -n_0 q_{n_0}(\tau). \tag{9.31}$$

Multiply both sides of Eqn (9.31) by $e^{n_0\tau}$ and use the product rule in differentiation. The result is that (9.29) can be expressed as

$$\frac{d}{d\tau}\left[e^{n_0\tau}q_{n_0}(\tau)\right] = 0.$$

Similarly multiply both sides of (9.29) by $e^{n(n_0+1-n)}$ and use the product rule again. Then Eqn (9.30) becomes

$$\frac{d}{d\tau}\left[e^{n(n_0+1-n)\tau}q_n(\tau)\right] = (n+1)(n_0-n)e^{n(n_0+1-n)\tau}q_{n+1}(\tau).$$

In these equations, let

$$u_n(\tau) = e^{n(n_0+1-n)\tau}q_n(\tau). \tag{9.32}$$

Then $u_n(\tau)$, $(n = 0, 1, 2, \ldots, n_0)$ satisfy the more convenient differential equations

$$\frac{du_{n_0}(\tau)}{d\tau} = 0, \tag{9.33}$$

$$\frac{du_n(\tau)}{d\tau} = (n+1)(n_0-n)e^{(2n-n_0)\tau}u_{n+1}(\tau). \tag{9.34}$$

In the simple epidemic only one infective is introduced so that the initial conditions $p_{n_0}(0) = 1$ and $p_n(0) = 0$ for $n = 0, 1, 2, \ldots, n_0-1$ translate, by (9.29) and (9.30), into

$$u_{n_0}(0) = 1 \text{ and } u_n(0) = 0 \text{ for } n = 0, 1, 2, \ldots, n_0 - 1. \tag{9.35}$$

The solution of the differential equation (9.33) is simply

$$u_{n_0}(\tau) = \text{ constant } = 1$$

using the initial condition in (9.35). From Eqn (9.34) the next equation in the iteration for $u_{n_0-1}(\tau)$ is

$$\frac{du_{n_0-1}(\tau)}{d\tau} = n_0 e^{(n_0-2)\tau}u_{n_0}(\tau) = n_0 e^{(n_0-2)\tau}.$$

This **separable differential equation** has the solution (obtained by direct integration)

$$u_{n_0-1}(\tau) = \int n_0 e^{(n_0-2)\tau}d\tau + C = \frac{n_0}{n_0-2}e^{(n_0-2)\tau} + C,$$

where C is a constant. Since $u_{n_0-1}(0) = 1$, then $C = -n_0/(n_0-2)$, so that

$$u_{n_0-1}(\tau) = \frac{n_0}{n_0-2}\left[e^{(n_0-2)\tau} - 1\right]. \tag{9.36}$$

Using this solution for $u_{n_0-1}(\tau)$, the equation for $u_{n_0-2}(\tau)$ from (9.33) is

$$\begin{aligned}
\frac{du_{n_0-2}(\tau)}{d\tau} &= 2(n_0-1)e^{(n_0-4)\tau}u_{n_0-1}(\tau) \\
&= 2(n_0-1)e^{(n_0-4)\tau}\frac{n_0}{n_0-2}\left[e^{(n_0-2)\tau} - 1\right] \\
&= \frac{2n_0(n_0-1)}{n_0-2}\left[e^{2(n_0-3)\tau} - e^{(n_0-4)\tau}\right].
\end{aligned} \tag{9.37}$$

This is also a separable differential equation, and routine integration together with the initial condition gives (for $n_0 > 4$)

$$u_{n_0-2}(\tau) = \frac{n_0(n_0-1)[(n_0-4)\{e^{2(n_0-3)\tau}-1\}-2(n_0-3)\{e^{(n_0-4)\tau}-1\}]}{(n_0-2)(n_0-3)(n_0-4)}.$$

This procedure can be continued but the number of terms increases at each step. The probabilities can be recovered from Eqns (9.29) and (9.31).

Example 9.10. *One infective is living with a group of four individuals who are susceptible to the disease. Treating this situation as a simple epidemic in an isolated group, find the probability that all members of the group have contracted the disease at time t.*

In the notation above we require $p_0(t)$. In the iteration scheme, $n_0 = 4$, so that from Eqns (9.32) and (9.33),

$$u_4(\tau) = 1, \qquad u_3(\tau) = 2(e^{2\tau}-1).$$

However, there is a change in (9.25) if $n_0 = 4$. The differential equation becomes

$$\frac{d}{d\tau}u_2(\tau) = 12(e^{2\tau}-1),$$

which, subject to $u_2(0) = 0$, has the solution

$$u_2(\tau) = -6(1+2\tau) + 6e^{2\tau}.$$

The equation for $u_1(\tau)$ is

$$\frac{d}{d\tau}u_1(\tau) = 36 - 36(1+2\tau)e^{-2\tau},$$

which has the solution

$$u_1(\tau) = 36(\tau-1) + 36(1+\tau)e^{-2\tau}.$$

A similar process gives the last function:

$$u_0(\tau) = 1 + (27-36\tau)e^{-4\tau} - (28+24\tau)e^{-6\tau}.$$

From Eqns (9.29) and (9.32) the probabilities can be now be found:

$$p_4(t) = e^{-4\beta t}, \quad p_3(t) = 2e^{-4\beta t} - e^{-6\beta t}, \quad p_2(t) = 6e^{-4\beta t} - (6+12\beta t)e^{-6\beta t},$$

$$p_1(t) = 36(\beta t-1)e^{-4\beta t} + 36(1+\beta t)e^{-6\beta t}, \quad p_0(t) = 1+9(3-4\beta t)e^{-4\beta t} - 4(7+6\beta t)e^{-6\beta t}.$$

A useful check on these probabilities is that they must satisfy

$$\sum_{n=0}^{4} p_n(t) = 1.$$

The formula for $p_0(t)$ gives the probability that all individuals have contracted the disease by time t. Note that $p_0(t) \to 1$ as $t \to \infty$, which means that all susceptibles ultimately catch the disease, a characteristic of the simple epidemic.

9.11 Problems

9.1. In a branching process the probability that any individual has j descendants is given by

$$p_0 = 0, \qquad p_j = \frac{1}{2^j}, \quad (j \geq 1).$$

Show that the probability generating function of the first generation is

$$G(s) = \frac{s}{2 - s}.$$

Find the further generating functions $G_2(s)$, $G_3(s)$, and $G_4(s)$. Show by induction that

$$G_n(s) = \frac{s}{2^n - (2^n - 1)s}.$$

Find the probability that the population size of the n-th generation is j given that the process starts with one individual. What is the mean population size of the n-th generation?

9.2. Suppose that in a branching process any individual has a probability given by the modified geometric distribution

$$p_j = (1 - p)p^j, \qquad (j = 0, 1, 2, \ldots),$$

of producing j descendants in the next generation, where p $(0 < p < 1)$ is a constant. Find the probability generating function for the second and third generations. What is the mean size of any generation?

9.3. A branching process has the probability generating function

$$G(s) = a + bs + (1 - a - b)s^2$$

for the descendants of any individual, where a and b satisfy the inequalities

$$0 < a < 1, \quad b > 0, \quad a + b < 1.$$

Given that the process starts with one individual, discuss the nature of the descendant generations. What is the maximum possible size of the n-th generation? Show that extinction in the population is certain if $2a + b \geq 1$.

9.4. A branching process starts with one individual. Subsequently any individual has a probability (Poisson)

$$p_j = \frac{\lambda^j e^{-\lambda}}{j!}, \qquad (j = 0, 1, 2, \ldots)$$

of producing j descendants. Find the probability generating function of this distribution. Obtain the mean and variance of the size of the n-th generation. Show that the probability of ultimate extinction is certain if $\lambda \leq 1$.

9.5. A branching process starts with one individual. Any individual has a probability

$$p_j = \frac{\lambda^{2j} \operatorname{sech} \lambda}{(2j)!}, \qquad (j = 0, 1, 2, \ldots)$$

of producing j descendants. Find the probability generating function of this distribution. Obtain the mean size of the n-th generation. Show that ultimate extinction is certain if λ is less than the computed value 2.065.

9.6. A branching process starts with two individuals. Either individual and any of their descendants has probability p_j, $(j = 0, 1, 2, \ldots)$ of producing j descendants independently of any other. Explain why the probabilities of $0, 1, 2, \ldots$ descendants in the first generation are

$$p_0^2, \quad p_0 p_1 + p_1 p_0, \quad p_0 p_2 + p_1 p_1 + p_2 p_0, \quad \cdots \quad , \sum_{i=0}^{n} p_i p_{n-i}, \ldots,$$

respectively. Now show that the probability generating function of the first generation is $G(s)^2$, where

$$G(s) = \sum_{j=0}^{\infty} p_j s^j.$$

The second generation from each original individual has generating function $G_2(s) = G(G(s))$ (see Section 9.2). Explain why the probability generating function of the second generation is $G_2(s)^2$, and of the n-th generation is $G_n(s)^2$.

If the branching process starts with r individuals, what would you think is the formula for the probability generating function of the n-th generation?

9.7. A branching process starts with two individuals as in the previous problem. The probabilities are given by

$$p_j = \frac{1}{2^{j+1}}, \quad (j = 0, 1, 2, \ldots).$$

Using the results from Example 9.1, find $H_n(s)$, the generating function of the n-th generation. Also find
(a) the probability that the size of the population of the n-th generation is $m \geq 2$;
(b) the probability of extinction by the n-th generation;
(c) the probability of ultimate extinction.

9.8. A branching process starts with r individuals, and each individual produces descendants with probability distribution $\{p_j\}$, $(j = 0, 1, 2, \ldots)$, which has the probability generating function $G(s)$. Given that the probability of the n-th generation is $[G_n(s)]^r$, where $G_n(s) = G(G(\ldots(G(s))\ldots))$, find the mean population size of the n-th generation in terms of $\mu = G'(1)$.

9.9. Let X_n be the population size of a branching process starting with one individual. Suppose that all individuals survive, and that

$$Z_n = 1 + X_1 + X_2 + \cdots + X_n$$

is the random variable representing the accumulated population size.
(a) If $H_n(s)$ is the probability generating function of the total accumulated population, Z_n, up to and including the n-th generation, show that

$$H_1(s) = sG(s), \qquad H_2(s) = sG(H_1(s)) = sG(sG(s))$$

(which perhaps gives a clue to the form of $H_n(s)$).
(b) What is the mean accumulated population size $\mathbf{E}(Z_n)$? (You do not require $H_n(s)$ for this formula.)
(c) If $\mu < 1$, what is $\lim_{n \to \infty} \mathbf{E}(Z_n)$, the ultimate expected population?
(d) What is the variance of Z_n?

9.10. A branching process starts with one individual and each individual has probability p_j of producing j descendants independently of every other individual. Find the mean and variance of $\{p_j\}$ in each of the following cases, and hence find the mean and variance of the population of the n-th generation:

(a)
$$p_j = \frac{e^{-\mu}\mu^j}{j!}, \qquad (j = 0, 1, 2, \ldots) \quad \text{(Poisson);}$$

(b)
$$p_j = (1-p)^{j-1}p \qquad (j = 1, 2, \ldots; 0 < p < 1) \quad \text{(geometric);}$$

(c)
$$p_j = \binom{r+j-1}{r-1} p^j (1-p)^r, \qquad (j = 0, 1, 2, \ldots; 0 < p < 1) \quad \text{(negative binomial)}$$

where r is a positive integer.

9.11. A branching process has a probability generating function

$$G(s) = \left(\frac{1-p}{1-ps}\right)^r, \qquad (0 < p < 1),$$

where r is a positive integer, the process being started with one individual (a negative binomial distribution). Show that extinction is not certain if $p > 1/(1+r)$. Find the probability of extinction if $r = 2$ and $p > \frac{1}{3}$.

9.12. Let $G_n(s)$ be the probability generating function of the population size of the n-th generation of a branching process. The probability that the population size is zero at the n-th generation is $G_n(0)$. What is the probability that the population becomes extinct at the n-th generation?
In Example 9.1, where $p_j = 1/2^{j+1}$ $(j = 0, 1, 2, \ldots)$, it was shown that

$$G_n(s) = \frac{n}{n+1} + \sum_{r=1}^{\infty} \frac{n^{r-1}}{(n+1)^{r+1}} s^r.$$

Find the probability of extinction,
(a) at the n-th generation,
(b) at the n-th generation or later.
What is the mean number of generations until extinction occurs?

9.13. An annual plant produces N seeds in a season which is assumed to have a Poisson distribution with parameter λ. Each seed has a probability p of germinating to create a new plant which propagates in the following year. Let M be the number of new plants. Show that p_m, the probability that there are m growing plants in the first year, is given by

$$p_m = (p\lambda)^m e^{-p\lambda}/m!, \quad (m = 0, 1, 2, \ldots),$$

that is, Poisson with parameter $p\lambda$. Show that its probability generating function is

$$G(s) = e^{p\lambda(s-1)}.$$

Assuming that all the germinated plants survive and that each propagates in the same manner in succeeding years, find the mean number of plants in year k. Show that extinction is certain if $p\lambda \leq 1$.

9.14. The version of Example 9.1 with a general geometric distribution is the branching process with $p_j = (1 - p)p^j$, $(0 < p < 1; \; j = 0, 1, 2, \ldots)$. Show that

$$G(s) = \frac{1-p}{1-ps}.$$

Using an induction method, prove that

$$G_n(s) = \frac{(1-p)[p^n - (1-p)^n - ps\{p^{n-1} - (1-p)^{n-1}\}]}{[p^{n+1} - (1-p)^{n+1} - ps\{p^n - (1-p)^n\}]}, \qquad (p \neq \tfrac{1}{2}).$$

Find the mean and variance of the population size of the n-th generation.

What is the probability of extinction by the n-th generation? Show that ultimate extinction is certain if $p < \tfrac{1}{2}$, but has probability $(1-p)/p$ if $p > \tfrac{1}{2}$.

9.15. A branching process starts with one individual, and the probability of producing j descendants has the distribution $\{p_j\}$, $(j = 0, 1, 2, \ldots)$. The same probability distribution applies independently to all descendants and their descendants. If X_n is the size of the n-th generation, show that

$$\mathbf{E}(X_n) \geq 1 - \mathbf{P}(X_n = 0).$$

In Section 9.3 it was shown that $\mathbf{E}(X_n) = \mu^n$, where $\mu = \mathbf{E}(X_1)$. Deduce that the probability of extinction eventually is certain if $\mu < 1$.

9.16. In a branching process starting with one individual, the probability that any individual has j descendants is $p_j = \alpha/2^j$, $(j = 0, 1, 2, \ldots, r)$, where α is a constant and r is fixed. This means that any individual can have a maximum of r descendants. Find α and the probability generating function $G(s)$ of the first generation. Show that the mean size of the n-th generation is

$$\mu_n = \left[\frac{2^{r+1} - 2 - r}{2^{r+1} - 1} \right]^n.$$

What is the probability of ultimate extinction?

9.17. Extend the tree in Figure 9.4 for the gambling martingale in Section 9.5 to Z_4, and confirm that

$$\mathbf{E}(Z_4 | Z_0, Z_1, Z_2, Z_3) = Z_3.$$

Confirm also that $\mathbf{E}(Z_4) = 1$.

9.18. A gambling game similar to the gambling martingale of Section 9.5 is played according to the following rules:
(a) the gambler starts with £1, but has unlimited resources;
(b) against the casino, which also has unlimited resources, the gambler plays a series of games in which the probability that the gambler wins is $1/p$ and loses is $(p-1)/p$, where $p > 1$;
(c) at the n-th game, the gambler either wins £$(p^n - p^{n-1})$ or loses £p^{n-1}.

If Z_n is the gambler's asset/debt at the n-th game, draw a tree diagram similar to that of Figure 9.3 as far as Z_3. Show that Z_3 has the outcomes

$$\{-p - p^2, -p^2, -p, 0, p^3 - p^2 - p, p^3 - p^2, p^3 - p, p^3\}$$

and confirm that

$$\mathbf{E}(Z_2 | Z_0, Z_1) = Z_1, \qquad \mathbf{E}(Z_3 | Z_0, Z_1, Z_2) = Z_2,$$

which indicates that this game is a martingale. Also show that

$$\mathbf{E}(Z_1) = \mathbf{E}(Z_2) = \mathbf{E}(Z_3) = 1.$$

Assuming that it is a martingale, show that, if the gambler first wins at the n-th game, then the gambler will have an asset gain or debt of $£(p^{n+1} - 2p^n + 1)/(p-1)$. Explain why a win for the gambler can only be guaranteed for all n if $p \geq 2$.

9.19. Let $X_1, X_2, \ldots$ be independent random variables with means $\mu_1, \mu_2, \ldots$ respectively. Let

$$Z_n = X_1 + X_2 + \cdots + X_n,$$

and let $Z_0 = X_0 = 0$. Show that the random variable

$$Y_n = Z_n - \sum_{i=1}^{n} \mu_i, \quad (n = 1,, 2, \ldots)$$

is a martingale with respect to $\{X_n\}$. [Note that $\mathbf{E}(Z_{n+1}|X_1, X_2, \ldots, X_n) = Z_n$.]

9.20. Consider an unsymmetric random walk ($p \neq \frac{1}{2}$) which starts at the origin. The walk advances one position with probability p and retreats one position with probability $1 - p$. Let X_n be the random variable giving the position of the walk at step n. Let Z_n be given by

$$Z_n = X_n + (1 - 2p)n.$$

Show that

$$\mathbf{E}(Z_2|X_0, X_1) = \{-2p, 2 - 2p\} = Z_1.$$

Generally show that $\{Z_n\}$ is a martingale with respect to $\{X_n\}$.

9.21. In the gambling martingale of Section 9.5, the random variable Z_n is the gambler's asset, in a game against a casino in which the gambler starts with £1 and doubles the bid at each play. The random variable Z_n has the possible outcomes

$$\{-2^n + 2m + 2\}, \quad (m = 0, 1, 2, \ldots, 2^n - 1).$$

Find the variance of Z_n. What is the variance of

$$\mathbf{E}(Z_n|Z_0, Z_1, \ldots, Z_{n-1})?$$

9.22. A random walk starts at the origin, and, with probability p_1, advances one position, and, with probability $q_1 = 1 - p_1$, retreats one position at every step. After 10 steps the probabilities change to p_2 and $q_2 = 1 - p_2$, respectively. What is the expected position of the walk after a total of 20 steps?

9.23. A symmetric random walk starts at the origin $x = 0$. The stopping rule that the walk ends when the position $x = 1$ is first reached is applied; that is, the stopping time T is given by

$$T = \min\{n : X_n = 1\},$$

where X_n is the position of the walk at step n. What is the expected value of T? If this walk was interpreted as a gambling problem in which the gambler starts with nothing with equal odds of winning or losing £1 at each play, what is the flaw in this stopping rule as a strategy of guaranteeing a win for the gambler in every game? Hint: the generating function for the probability of the first passage is

$$G(s) = [1 - (1 - s^2)^{\frac{1}{2}}]/s :$$

see Problem 3.11.

9.24. In a finite-state branching process, the descendant probabilities are, for every individual,

$$p_j = \frac{2^{m-j}}{2^{m+1} - 1}, \qquad (j = 0, 1, 2, \ldots, m),$$

and the process starts with one individual. Find the mean size of the first generation. If X_n is the size of the n-th generation, explain why

$$Z_n = \left[\frac{2^{m+1} - 1}{2^{m+1} - m - 2}\right]^n X_n$$

defines a martingale over $\{X_n\}$.

9.25. A random walk starts at the origin, and at each step the walk advances one position with probability p or retreats with probability $1 - p$. Show that the random variable

$$Y_n = X_n^2 + 2(1 - 2p)nX_n + [(2p - 1)^2 - 1]n + (2p - 1)^2n^2,$$

where X_n is the random variable of the position of the walk at time n, defines a martingale with respect to $\{X_n\}$.

9.26. A simple epidemic has n_0 susceptibles and one infective at time $t = 0$. If $p_n(t)$ is the probability that there are n susceptibles at time t, it was shown in Section 9.7 that $p_n(t)$ satisfies the differential-difference equations (see Eqns (9.15 and (9.16))

$$\frac{dp_n(t)}{dt} = \beta(n + 1)(n_0 - n)p_{n+1}(t) - \beta n(n_0 + 1 - n)p_n(t),$$

for $n = 0, 1, 2, \ldots n_0$. Show that the probability generating function

$$G(s, t) = \sum_{n=0}^{n_0} p_n(t)s^n$$

satisfies the partial differential equation

$$\frac{\partial G}{\partial t} = \beta(1 - s)\left[n_0\frac{\partial G}{\partial s} - s\frac{\partial^2 G}{\partial s^2}\right]$$

(see Bailey (1964), Ch.12).

Nondimensionalise the equation by putting $\tau = \beta t$. For small τ let

$$H(s, \tau) = G(s, \tau/\beta) = H_0(s) + H_1(s)\tau + H_2(s)\tau^2 + \cdots.$$

Show that

$$nH_n(s) = n_0(1 - s)\frac{\partial H_{n-1}(s)}{\partial s} - s(1 - s)\frac{\partial^2 H_{n-1}(s)}{\partial s^2},$$

for $n = 1, 2, 3, \ldots n_0$. What is $H_0(s)$? Find the coefficients $H_1(s)$ and $H_2(s)$. Now show that the mean number of infectives for small τ is given by

$$n_0 - n_0\tau - \tfrac{1}{2}n_0(n_0 - 2)\tau^2 + O(\tau^3).$$

CHAPTER 10

Brownian Motion: Wiener Process

10.1 Introduction

In this chapter we shall consider processes continuous in time, and with continuous state spaces. In previous chapters we have discussed processes where both stage and state are discrete: random walks (Chapter 3) where the stage is the number of steps and the state is the position of the walk and where the stage is continuous and the state is discrete: for example, birth processes (Chapter 6) where the stage is time and the state is the number of births. Note that unlike birth processes there is no possibility of an "event" or movement not occurring in any time interval. The processes developed here may be defined in terms of a continuous random variable $X(t)$ which depends on continuous time $t > 0$, although extension to more than a one-dimensional random variable is possible, and as shown below has been useful in modelling certain physical behaviours. However, we shall be explaining the one-dimensional case in detail.

10.2 Brownian motion

Brownian motion is named after Robert Brown[1], who was one of the first to comment on the phenomenon when observing through a microscope the seemingly random and continuous movement of pollen grains suspended in water. This, of course is a process in three dimensions and would require modelling using three random variables—but more of this later. This stochastic process was used by Einstein[2] in the so-called "miraculous year" of 1905. This was the year when he postulated the molecular or atomic hypothesis on the movement or continuous motion of particles caused by collisions with molecules in a fluid in which they were suspended[3]. This was also the year when he propounded the special theory of relativity and when his work on the photoelectric effect was carried out, and which in 1922 was awarded the Nobel prize in Physics.

---

[1] Robert Brown (1773–1858), Scottish botanist.

[2] Albert Einstein (1879–1955), German theoretical physicist.

[3] 'In this paper it will be shown that, according to the molecular theory of heat, bodies of a microscopically visible size suspended in liquids must as a result of thermal molecular motions, perform motions of such magnitudes that they can be easily observed with a microscope. It is possible that the motions to be discussed here are identical with so-called Brownian molecular motion; however, the data available to me on the latter are so imprecise that I could not form a judgement on the question...' Einstein (1956 from 1905 paper) regarding

Two-dimensional Brownian motion can be used, for example, to model spatial movement over time of insects on a plot of land; one-dimensional Brownian motion has been used in finance to track stock movements. Bachelier[4] carried out early work in the mathematical modelling of Brownian motion, and went on to apply it to problems in this area. The tracking of stock movements over time is an example of an approximation by a continuous process of an essentially discrete process observed over a long period at small intervals. When a plot of realisations of $X(t)$ at discrete time points is viewed over decreasing intervals of time it appears to be continuous. The idea of Brownian motion as a limit of a discrete process such as a random walk (Chapter 3) is discussed in the next section.

Consider the continuous random variable $X(t)$ which is a function of time t and suppose that $X(0) = 0$ such that for any time interval (t_i, t_{i+1}), the increments or changes

$$D(t_i, t_{i+1}) = [X(t_{i+1}) - X(t_i)] \qquad (10.1)$$

over n non-overlapping time periods are mutually independent and depend only on

$$X(t_i), \qquad t_i < t_{i+1}, \quad 0 \le i \le n,$$

where n is finite: this is a **Markov property**. Furthermore it is assumed that the $D(t_i, t_{i+1})$ have the same probability distribution for $0 \le i \le n$ which depend only on the length of the interval (t_i, t_{i+1}): in other word the process possesses **stationary independent increments**.

To simplify matters it is assumed that this is a process with the expected value $\mathbf{E}[X(t)] = 0$: this defines a process with **no drift**. The random variable $X(t)$ is, in fact, $D(0, t)$ so that $X(s)$ will have the same probability distribution as $D(t, t + s)$ for all s. A consequence of all these assumptions is that $X(t)$ can be shown to have a normal distribution. Generally a stochastic process with stationary, independent increments is called a **Lévy process** [5]: hence Brownian motion is such a process. The name **Wiener process**[6] is also now more common when discussing technical details. The term Brownian motion is often used for the phenomenon of diffusion but both seem to be used interchangeably for the process: we use both terms here.

To summarise, a **Wiener process** or **Brownian motion** for a continuous random variable $X(t)$ has the following properties:

(a) $X(0) = 0$;

(b) $D(t_i, t_{i+1}) = [X(t_{i+1}) - X(t_i)]$ has a normal distribution with mean 0 and variance $\sigma^2(t_{i+1} - t_i)$, $1 \le i \le n - 1$;

(c) $D(t_i, t_{i+1})$ are mutually independent in non-overlapping time intervals.

Note that, by choosing $t_1 = 0$ and $t_2 = 1$, then $\sigma^2 = \mathbf{V}[X(1)]$.

The process where $\sigma^2 = 1$ is known as **standard Brownian motion** or Wiener process. Any (non-standard) process can be converted to standard process $W(t)$ by defining $W(t) = X(t)/\sigma$.

---

[4] Louis Bachelier (1870–1946), French mathematician.
[5] Paul Lévy (1886–1971), French mathematician.
[6] Norbert Wiener (1894–1954), American mathematician.

The dispersion of standard Brownian motion can be shown by viewing computed outputs against the variance of the normal distribution. The probability density function of standard Brownian motion is given by

$$f_t(x) = \frac{1}{\sqrt{2\pi t}} \exp\left(\frac{-x^2}{2t}\right), \quad (-\infty < x < \infty), \quad (t > 0),. \tag{10.2}$$

Hence the variance of this normal distribution is $\mathbf{V}(X) = t$. Figure 10.1 shows graphs of five computed Brownian motions together with a measure of dispersion given by the standard deviation shown by the curve $x = \sqrt{t}$. Several trials of Brownian motion indicate that any particular output can diverge considerably.

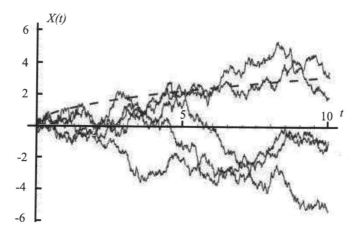

Figure 10.1 *Includes five Brownian motions and the dashed curve representing the standard deviation* $x = \sqrt{t}$.

Wiener processes are part of a larger class of processes known as **diffusion processes**, and following this the variance σ^2 is also known as the diffusion coefficient.

Condition (a) may be relaxed to include processes starting at a fixed point $X(0) = x$ rather than at the origin, but these translated processes are still Wiener processes.

10.3 Wiener process as a limit of a random walk

Let us return to the idea of illustrating a Wiener process as the limit of a random walk. Recall from Chapter 3 the description of the *symmetric* one-dimensional unrestricted random walk with step length ± 1 (see Section 3.2). Since it is symmetric, or in the terminology above, the walk has no drift, the probability of a positive or negative step is $\frac{1}{2}$, and the position of the walk, X_n, at stage n has a distribution with $\mathbf{E}(X_n) = 0$ and $\mathbf{V}(X_n) = n$.

We adapt the symmetric random walk in the following way to create a limiting process on the walk. Divide the time-scale $t \in [0, \infty)$ into intervals of fixed (small) length ε. At each step $t = 0, \varepsilon, 2\varepsilon, \ldots$ there is a probability $\frac{1}{2}$ that the walk advances $\sqrt{\varepsilon}$ or a probability $\frac{1}{2}$ that the walk retreats $-\sqrt{\varepsilon}$. The limiting process is such that

$\varepsilon \to 0$. Let Q_i be the outcome of the i-th step (that is at time $i\varepsilon$) so that

$$Q_i = \begin{cases} \sqrt{\varepsilon} & \text{probability } \frac{1}{2} \\ -\sqrt{\varepsilon} & \text{probability } \frac{1}{2} \end{cases} \tag{10.3}$$

for $i = 0, 1, 2, \ldots$ or at $t = 0, \varepsilon, 2\varepsilon, \ldots$ (the reason for $\sqrt{\varepsilon}$ will be explained shortly). If X_i is the random variable of the position of the walk at step i, then

$$X_i = X_0 + \sum_{j=1}^{i} Q_j = \sum_{j=1}^{i} Q_j, \tag{10.4}$$

assuming $X(0) = 0$. Now

$$\mathbf{E}(Q_n) = \frac{1}{2}\sqrt{\varepsilon} - \frac{1}{2}\sqrt{\varepsilon} = 0, \qquad \mathbf{V}(Q_n) = \mathbf{E}(Q_n^2) - \mathbf{E}(Q_n)^2 = \varepsilon, \tag{10.5}$$

for all n. Suppose that we consider a fixed interval of time $[0, n\varepsilon]$. Over $n\varepsilon$,

$$\mathbf{E}(X_n) = 0, \qquad \mathbf{V}(X_n) = \mathbf{V}\left[\sum_{j=1}^{n} Q_j\right] = n\mathbf{V}(Q_n) = n\varepsilon. \tag{10.6}$$

We have to show that (b) and (c) hold. Consider any two times $t_1 = i_1\varepsilon$ and $t_2 = i_2\varepsilon$ where $t_2 > t_1$. Then

$$D(t_1, t_2) = X(t_2) - X(t_1) = \sum_{j=i_1+1}^{i_2} Q_j, \tag{10.7}$$

which only depends on the number of outcomes in the interval $[t_1, t_2]$ that establishes (c).

The expectation

$$\mathbf{E}[X(t_2) - X(t_1)] = \mathbf{E}\left[\sum_{j=i_1+1}^{n_i} Q_j\right] = \sum_{j=i_1+1}^{i_2} \mathbf{E}(Q_j) = 0, \tag{10.8}$$

and the variance

$$\mathbf{V}[X(t_2) - X(t_1)] = \sum_{j=i_1+1}^{i_2} \mathbf{V}(Q_j) = (i_2 - i_1)\varepsilon = t_2 - t_1. \tag{10.9}$$

The random variables $D(t_1, t_2) = X(t_2) - X(t_1)$ have variances that depend only on the interval length with $\sigma^2 = 1$, *which is the reason why the step length $\sqrt{\varepsilon}$ was chosen.* This means that this is a standard Brownian motion.

An example of computed Brownian motion is shown in Figure 10.2. A feature of Brownian motion is that a magnification of a section of a trajectory looks like the original.

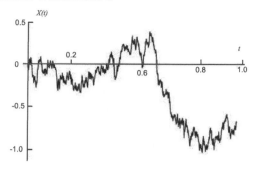

Figure 10.2 *An example of a standard Brownian motion over the interval $t \in [0, 1]$ (this was computed using a Wiener process package in* Mathematica *over 1,000 time-points).*

10.4 Brownian motion with drift

Consider the random variable

$$U(t) = X(t) + \mu t, \tag{10.10}$$

where $X(t)$ is the continuous random variable defining a Brownian motion of Section 10.2 with $X(0) = 0$ and variance $\sigma^2 t$: the constant $\mu \neq 0$ is known as the **drift** of the process. Obviously $U(0) = 0$ and using properties of expectation and variance that

$$\mathbf{E}[U(t)] = \mu t, \qquad \mathbf{V}[U(t)] = \mathbf{V}[X(t) + \mu t] = \mathbf{V}[X(t)] = \sigma^2 t, \tag{10.11}$$

by the variance rule (1.12). In this case the increment

$$D_u(t_i, t_{i+1}) = U(t_{i+1}) - U(t_i) - D(t_i, t_{i+1}) + \mu(t_{i+1} - t_i) \tag{10.12}$$

by (10.1). Its variance is given by

$$
\begin{aligned}
\mathbf{V}[D_u(t_i, t_{i+1})] &= \mathbf{V}[U(t_{i+1}) - U(t_i)] \\
&= \mathbf{V}[D(t_i, t_{i+1}) + \mu(t_{i+1} - t_i)] \\
&= \mathbf{V}[D(t_i, t_{i+1})],
\end{aligned} \tag{10.13}
$$

using the variance rule (1.12) again. Hence $U(t)$ meets the requirements for a Wiener process except that its mean is displaced. The arguments in terms of the random walk of the previous section apply similarly to Brownian motion with drift.

A realisation (or sample path) of a Brownian motion with drift is shown in Figure 10.3.

10.5 Scaling

We mentioned earlier the realisations of Brownian motion and the approximation of a continuous process by observing $X(t)$ at smaller intervals of time. We will address this more formally through the introduction of the idea of scaling both the vertical (position) and horizontal (time) axes of plots of the process which naturally affect their smoothness.

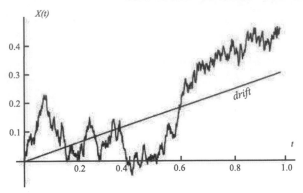

Figure 10.3 *An example of Brownian motion with drift* $\mu = 0.3$ *and volatility* 0.3 *computed over over 1,000 time points on the interval* $t \in [0, 1]$.

Suppose that the random variable $X(t)$ represents a one-dimensional Brownian motion with $X(0) = 0$ and no drift. We have already encountered the notion of scaling the vertical axis by choosing to observe the standard Brownian motion $W(t) = X(t)/\sigma$ where the smoothness of a resulting plot would depend on whether σ^2 is greater than or less than 1. Consider now the transformation of $X(t)$ to $S(t)$ given by

$$S(t) = aX(t/b), \tag{10.14}$$

where $a \neq 0$ and $b > 0$ are constants. Clearly $S(t)$ is a linear function of a continuous normally distributed random variable $X(t)$, and so is also normally distributed but with mean 0 and variance $a^2\sigma^2 t/b$. Hence the properties of the increments of $S(t)$ will satisfy the conditions for it to be a Brownian motion.

A particular case of this transformation occurs when $a = -1$ and $b = 1$, which results in the process $-X(t)$ being a Brownian motion. This is known as the **reflection** of the original process, with the same properties as the original. This is perhaps not surprising given that both positive and negative increments are possible in the original process. The use of the reflected process enables us to investigate whether a realisation does or does not return to the origin in a specified time interval known as the **reflection principle** (first introduced in Problem 3.9 in the chapter 3 on random walks).

Realisations or **sample paths** are never smooth no matter how small the time interval over which they are observed. An illustration of this may be obtained by considering the particular transformation $a = \sqrt{c}$ and $b = c$, where c is a positive constant, giving rise to the Brownian motion $Q(t) = \sqrt{c}X(t/c)$ whose *change* has mean 0 and variance $\sigma^2 t$. In other words the distributional properties of $Q(t)$ are the same as those of $X(t)$ whatever the value of the constant c. It does not matter how much the two axes are stretched given increasing detail or *granularity* of the path: $Q(t)$ still acts in a similar way to the original path. However, for small time intervals, realisations of the path will always be *spiky*. This parallels a phenomenon known as a **fractal**.

A fractal is a mathematical set that shows a repeating pattern which is visible at every scale no matter how small: in other words, if the pattern is magnified it looks the same. See Addison (1997) for explanation concerning fractals and their relation to Brownian motion. It is also known as a self-similar pattern. The **Mandelbrot set** is a particular fractal named after its creator[7]. However, a fractal is usually defined deterministically by a mathematical formula, unlike Brownian motion, which is generated probabilistically. In a similar way any snapshot of Brownian motion for different scalings will be statistically similar. A sample path of Brownian motion can be shown to be **almost surely continuous**[8].

Consider the increment in a standard Brownian motion,

$$D_s(t, t + \delta) = W(t + \delta) - W(t) : \tag{10.15}$$

it will be normally distributed with mean 0 and variance δ. Hence $[D_s(t, t + \delta)]^2/\delta$ has a χ^2 distribution on one degree of freedom which has mean 1 and variance 2, from which it follows that $[D_s(t, t + \delta)]^2$ has mean δ and variance 2δ.

Following Lawler (2005), for small δ. the approximate size of the increment is therefore $\pm\sqrt{\delta}$ since the variance is small and both tend to zero. The purpose of this diversion is to establish intuitively the nature of the output $W(t)$. Is $W(t)$ against t a smooth curve? To answer this question we need to examine the ratio

$$\frac{W(t + \delta) - W(t)}{\delta} = \frac{D_s(t, t + \delta)}{\delta}$$

as $\delta \to 0$. The mean of $D_s(t, t + \delta)$ is of order $\sqrt{\delta}$, which implies that

$$\lim_{\delta \to 0} \frac{W(t + \delta) - W(t)}{\delta} = O(1/\sqrt{\delta}),$$

which implies that the limit does not exist. Hence $W(t)$ is nowhere differentiable.

10.6 First visit times

Consider the one-dimensional Brownian motion, $X(t)$, with no drift, which has mean 0 and variance $\sigma^2 t$, starting at the origin. In this section, we are interested in the distribution of the time T_a for $X(t)$ to reach a value $a(\neq 0)$ for the first time: also known as the first passage time to a. It can be shown that the Brownian motion is **recurrent** (see Lawler (2006) or Grimmett and Stirzaker (1982)).

To obtain the probability distribution of T_a we need to determine $\mathbf{P}(T_a < t) = \mathbf{P}(A)$, where A is the event $T_a < t$, that is, a is reached before time t, which will depend on the random variable $X(t)$. To derive this result it is important to note that the path is continuous despite the the fact that *realisations* or plots are based on evaluation at discrete times. The derivation of $\mathbf{P}(A)$ is based on what happens to the

---

[7] Benoit Mandelbrot (1924–2010), Polish mathematician.

[8] *Almost surely continuity* in probability means that the process is continuous with probability 1. We have avoided undue mathematical rigour but this term arises from the difficulty in defining this probability. For example, the certainty can arise over events which are certain but over an infinite time which can lead to logical inconsistencies.

path after a is reached. Suppose that the event $B = (X(t) > a)$, then by the law of total probability,

$$\mathbf{P}(A) = \mathbf{P}(A \cap B) + \mathbf{P}(A \cap B^c), \tag{10.16}$$

where B^c is the event $X(t) \leq a$. The right-hand side of the previous equation may be written as

$$\mathbf{P}(A|B)\mathbf{P}(B) + \mathbf{P}(B^c|A)\mathbf{P}(A). \tag{10.17}$$

If the event B^c (that is, event $X(t) \leq a$) given A has occurred then it is equally likely that B has occurred given A. This is the **reflection principle** due to André[9] (see Figure (10.4)). Hence

$$\mathbf{P}(B^c|A) = \tfrac{1}{2}. \tag{10.18}$$

Furthermore, $\mathbf{P}(A|B) = 1$ since if $X(t) > a$ then $T_a < t$. Hence from (10.17),

$$\mathbf{P}(A) = \mathbf{P}(B) + \tfrac{1}{2}\mathbf{P}(A),$$

so that

$$\mathbf{P}(A) = 2\mathbf{P}(B). \tag{10.19}$$

In words, this interesting result states that:

in a Brownian motion starting from the origin, the probability that level a is reached first within time t is twice the probability that the random variable $X(t)$ is greater than a.

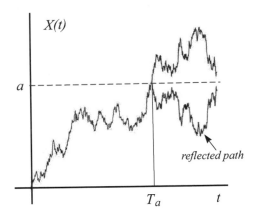

Figure 10.4 *Illustrating the reflection principle: T_a is the time to the first visit to level a.*

Since $X(t)$ has a normal distribution with mean 0 and variance $\sigma^2 t$, it follows that the distribution function of T_a is given by

$$\mathbf{P}(T_a < t) = 2[1 - \Phi(z)], \tag{10.20}$$

where Φ is a standard normal distribution function and $z = a/(\sigma\sqrt{t})$: this is known

---

[9] Désiré André (1840–1917), French mathematician.

as the **inverse normal distribution function** and the density function is

$$\frac{a}{\sqrt{2\pi\sigma^2 t^3}}\exp\left[-\frac{a^2}{2\sigma^2 t}\right], \qquad t > 0. \tag{10.21}$$

Figure 10.4 illustrates the reflection principle which is the basis of (10.20).

10.7 Other Brownian motions in one dimension

We have already met linear transformations of $X(t)$ to standard Brownian motion and Brownian motion with drift. In this section we shall discuss processes where observations over time do not satisfy the Wiener conditions for Brownian motion but may be transformed to satisfy the conditions.

Geometric Brownian motion. This process is often used to model financial stock prices or population growth, or in other situations where measurements cannot be negative. In finance it forms the basis for the Black–Scholes[10] equation where the log returns of stock prices is modelled: the main interest is in the option pricing of derivatives[11].

Geometric Brownian motion $G(t)$ is defined by

$$G(t) = G(0)\exp[(\mu - \tfrac{1}{2}\sigma^2)t + \sigma X(t], \quad (t \geq 0), \tag{10.22}$$

where $X(t)$ is a standard Brownian motion and $G(0) > 0$ is the initial value of the process. In this formula, μ is the drift which is defined in finance by a deterministic or predictable long-term trend, and σ measures variation and unpredictability, which also might have a trend given by the term $\tfrac{1}{2}\sigma^2 t$. The parameter σ is sometimes known as the **volatility** of the market in finance. Note that $G(t) \geq 0$, which is required for financial data.

The random variable $G(t)$ is not itself a Brownian motion, but

$$\log[G(t)] = \log[G(0)] + (\mu - \tfrac{1}{2}\sigma^2)t + \sigma X(t)$$

is, since it is essentially a Brownian motion with drift. The process $G(t)$ is said to have a **lognormal distribution**. A transformation to give the path in the original measurements is possible by taking the exponential of the Brownian motion. An illustration of a geometric Brownian motion is shown in Figure 10.5. The probability density function of $G(t)$ is

$$f_t(x) = \frac{1}{\sigma x \sqrt{2\pi t}}\exp\left[-\frac{[\log x - \log G(0) - (\mu - \tfrac{1}{2}\sigma^2)t]^2}{2\sigma^2 t}\right]. \tag{10.23}$$

From the density function it can be shown that the mean and variance of $G(t)$ are given by

$$\mathbf{E}[G(t)] = G(0)e^{\mu t}, \tag{10.24}$$

---

[10] Fischer Black (1938–1995), Myron Scholes (1941–), American economists.

[11] A *derivative* is an traded asset with a price that is dependent upon or derived from one or more underlying assets which can include shares, bonds, commodities, currencies, interest rates, and market indexes.

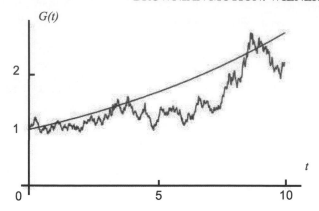

Figure 10.5 *Geometric Brownian motion G(t) with $\mu = 0.1$, $\sigma = 0.2$, and initial value $G(0) = 1$ drawn using 1,000 time-points. The smooth curve shows the mean $\mathbf{E}[G(t)] = G(0)e^{\mu t}$, known also as the* **trend**.

$$\mathbf{V}[G(t)] = G^2(0)e^{2\mu t}(e^{\sigma^2 t} - 1). \tag{10.25}$$

The result for the exponential growth in the mean explains why the term $\frac{1}{2}\sigma^2 t$ is present in $G(t)$.

Some examples of probability densities for standard Brownian, motion with drift and geometric Brownian motion are shown in Figure 10.6.

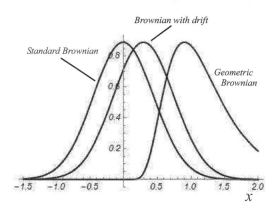

Figure 10.6 *Sample density functions for standard Brownian motion (mean 0, volatility 1), Brownian motion with drift (drift 1.5, volatility 1), geometric Brownian motion (mean 1, volatility 1, initial value $G(0) = 1$), all computed in the section $t = 1$.*

Ornstein–Uhlenbeck process[12] This process is concerned with the modelling of a particle moving through a gas, for example, subject to a frictional drag and the random bombardment of molecules of the gas. The resulting Brownian motion depends on three parameters: the mean μ, velocity λ, and σ^2.

[12] Leonard Ornstein (1880–1941), George Eugene Uhlenbeck (1900–1988) Dutch physicists.

In finance this is termed a **mean reverting process**. For example, it has applications in commodity pricing where economic arguments suggest that prices increase to a long-term mean value when they are considered too low and decrease when considered too high. The derivation of results for this process involves solution of **stochastic differential equations**, which is beyond the scope of this text: readers are referred to Grimmett and Stirzaker (1982) for further details.

10.8 Brownian motion in more than one dimension

Brownian motion in two or three dimensions[13] is a more likely phenomenon than

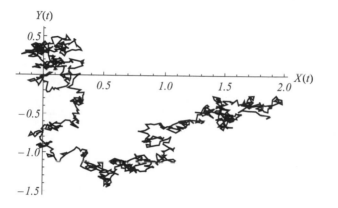

Figure 10.7 *Two-dimensional standard Brownian motion starting at the origin with 1,000 time points in both directions.*

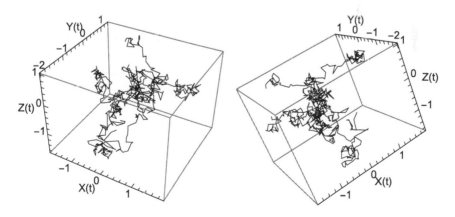

Figure 10.8 *Three-dimensional standard motion starting at the origin with 100 time points in each direction: there are two viewpoints of the same Brownian motion.*

the one-dimensional case. In the opening section we considered the example of the

[13] These were computed using a standard *Mathematica* package

motion of a particle in a fluid in three or two dimensions. The modelling of these situations is relatively straightforward when variables are independent with or without drift. The data for two-dimensional standard Brownian motion are generated in simultaneous pairs to create a consecutive sequence of points in the plane. A similar process generates three-dimensional Brownian motion. It is fairly easy to show computed diagrams of Brownian motion. Illustrations of standard Brownian in two and three dimensions are shown in Figures 10.7 and 10.8.

10.9 Problems

10.1. Let $X(t)$ be a standard Brownian motion.
  (a) Find $\mathbf{P}[X(2) > 3]$.
  (b) Find $\mathbf{P}[X(3) > X(2)]$.

10.2 Let $X(t)$ be a Brownian motion with mean 0 and variance $\sigma^2 t$ starting at the origin.
  (a) Find the distribution of $|X(t)|$, the absolute distance of $X(t)$ from the origin.
  (b) If $\sigma = 1$, evaluate $\mathbf{P}(|X(5)|) > 1)$.

10.3. $X(t) = \ln[Z(t)]$ is a Brownian motion with variance $\sigma^2 t$.
  (a) Find the distribution of $Z(t)$.
  (b) Evaluate $\mathbf{P}[Z(t) > 2]$ when $\sigma^2 = 0.5$.

10.4. Let $X(t), (t \geq 0)$ be a standard Brownian motion. Let τ_a be stopping time for $X(t)$ (that is, the first time that the process reaches state a). Explain why

$$Y(t) = \begin{cases} X(t) & 0 < t < \tau_a, \\ 2X(\tau_a) - X(t) & t \geq \tau_a \end{cases}$$

represents the reflected process. Show that $Y(t)$ is also a standard Brownian motion.

10.5. For a standard Brownian motion $X(t), (t \geq 0)$, show that $\mathbf{E}[X^2(t)] = t$ using the mgf for $X(t)$.

10.6. In a standard Brownian motion let t_a be the first time that the process reaches $a \geq 0$, often known as the **hitting time**. Let $Y(t)$ be the maximum value of $X(s)$ in $0 \leq s \leq t$. Both t_a and $Y(t)$ are random variables with the property that $t_a \leq t$ if and only if $Y(t) \geq a$. Using the reflection principle (Section 10.7), we know that

$$\mathbf{P}[Y(t) \geq a | t_a \leq t] = \tfrac{1}{2}.$$

Using this result, show that

$$\mathbf{P}(t_a \leq t) = \sqrt{\left(\frac{2}{\pi t}\right)} \int_a^\infty \exp[-x^2/(2t)]dx.$$

What is the mean hitting time?

10.7. $X(t)$ is a standard Brownian motion. Show that $Y(t) = X(a^2 t)/a$ where $a > 0$ is a constant is also a standard Brownian motion.

10.8. $X(t)$ and $Y(t)$ are independent standard Brownian motions.
  (a) Find the probability densities of $U(t) = X^2(t)/t$ and $V(t) = Y^2(t)/t$.

(b) Using mgf's (see Problem 1.29), show that the probability distribution of $W^2(t) = U^2(t) + V^2(t)$ is exponential.

(c) Find $\mathbf{P}[R(t) \geq r]$ where $R(t) = \sqrt{[X^2(t) + Y^2(t)]}$ is the **Euclidean distance** of the two-dimensional Brownian motion $[X(t), Y(t)]$ from the origin. (See Open University (1988), Unit 14: *Diffusion Processes*.)

10.9. $X(t)$ is a standard Brownian motion. Find the moment generating function of

$$Y(t) = \begin{cases} 0 & t = 0, \\ tX(1/t) & t > 0. \end{cases}$$

Show that

$$\lim_{t \to \infty} \frac{X(t)}{t} = 0.$$

10.10. The probability density function for geometric Brownian motion is given by

$$f_x(t) = \frac{1}{\sigma x \sqrt{2\pi t}} \exp[-(\ln x - (\mu - \tfrac{1}{2}\sigma^2)t)^2/(2\sigma^2 t)], \quad (x \geq 0)$$

for the random variable

$$Z(t) = \exp[(\mu - \frac{1}{2}\sigma^2)t + \sigma X(t)], \quad (x > 0).$$

Show that the mean and variance of this lognormal distribution are given by

$$\mathbf{E}[Z(t)] = e^{\mu t}, \qquad \mathbf{V}[Z(t)] = e^{2\mu t}[e^{\sigma^2 t} - 1].$$

10.11. If $X(t)$, $(t \geq 0)$ is a standard Brownian motion, use the conditional expectation

$$\mathbf{E}[X(t)|X(u), 0 \leq u < y], \qquad (0 \leq y < t)$$

to show that $X(t)$ is a martingale (see Section 9.5 for definition of martingales).

CHAPTER 11

Computer Simulations and Projects

The software *Mathematica* has been used extensively by the authors to confirm answers in worked examples and in the problems at the end of chapters, and in generating graphs. Other symbolic computation systems such as $Maple^{TM}$, $MATLAB^{TM}$, etc., could also have been used since they are frequently part of many courses in mathematics, statistics, engineering, and the physical sciences. It is seen as a necessary training in the use of computers in these subjects.

Although the symbolic computational systems may be used to solve numerical problems alternative statistical software such as R: R can be downloaded from

http://www.r-project.org

Any statistical software with a suite of simulation routines that allow the generation of results for the probabilistic questions and procedures for the manipulation of matrices in Markov chains, as well as graphical routines, can be used for the numerical projects.

Readers of this book are encouraged to sample some of the theoretical and numerical projects listed below in conjunction with the text. The following projects follow the chapters in the main text. It should be emphasised that any computer routine will generally generate answers without explaining just how any results were obtained. Symbolic computing is not yet a substitute for understanding the theory behind stochastic processes. Computations and graphs are often interesting in that they give some idea of scale and importance in problems: a result can look interesting as a formula but be of negligible consequence numerically as a model of a stochastic process.

The programs can be considered as a supplement to the text for those readers who have access to *Mathematica* and R.

Chapter 1: Some Background on Probability

11-1.1. Two distinguishable dice are rolled. The program simulates the sum of the two face values shown for 1,000, say, random throws. Compare the average of these values with the theoretical expected sum of 7 as in Problem 1.11. Compare also theoretical variance of the sum with the simulated value.

11-1.2. The computer program simulates the case of three dice which are rolled and their face values noted. By running the program, say, 2,000 times, compare by calculating the ratio of the number of the cases in which two dice show the same face value and the third is different with the theoretical probability of $5/12$. Estimate the probability that the three dice all show different values with the theoretical probability.

11-1.3. A fair coin is spun n times, and the sequence of faces shown recorded. The program simulates this Bernoulli experiment (see Section 1.7). Count the frequencies of the occurrences of heads and tails over, say, $n = 1,000$ trials.

Continue the program by counting in a simulation the number of times, say, a head appears in say 20 spins, and perform the simulation $m = 2,000$ times. Compare the data obtained with the exact binomial distribution which the theory predicts for this problem.

11-1.4. The gamma distribution has the density function

$$f(x) = \frac{\alpha^n}{\Gamma(n)} x^{n-1} e^{-\alpha x}.$$

The program shows graphically by surfaces the distribution and density functions for $n = 2$, say, in terms of x and the parameter $\lambda = 1/\alpha$ (*Mathematica* uses the parameter λ).

11-1.5. This program displays the moment generating function for Z which has the normal distribution $N(0, 1)$, and also for Z^2 (see Problem 1.29).

Chapter 2: Some Gambling Problems

11-2.1. This program simulates the gambler's ruin problem in which a gambler with k units bets against an opponent with $(a - k)$ with a probability p of the gambler winning at each play. It shows a graph of the gambler's current stake against the number of the play, and such that the graph terminates when either of the absorbing states at $k = 0$ or $k = a$ is reached. See Figure 2.1 in Section 2.3.

11-2.2. In a single trial the possible score outcomes 1 and 2 can occur with probabilities α and $1 - \alpha$ where $0 < \alpha < 1$ (see Problem 2.10). A succession of such trials take place and each time the scores are accumulated from zero. The program simulates this process, and counts the number of times the total scores 1,2,3,..., m occur. Possible numbers are $\alpha = \frac{1}{4}, m = 20$ scores run 1,000 times. The theory gives the probability that score n occurs as

$$p_n = \frac{1 - (\alpha - 1)^{n+1}}{2 - \alpha}.$$

Compare the results graphically.

11-2.3. The probability of ruin in the standard gambler's ruin problem is given by

$$u_k = \frac{s^k - s^a}{1 - s^a}, \quad (s \neq 1), \quad u_k = 1 - \frac{k}{a}, \quad (s = 1),$$

where $s = (1 - p)/p$, a is the total stake, k is the gambler's initial stake, and p is the probability that the gambler wins at each play. The program reproduces Figure 2.2 which shows the probability u_k versus k for a given value of a ($a = 20$ in the figure) for a selection of values of p.

11-2.4. The expected duration of the gambler's ruin is given by

$$d_k = \frac{1}{1 - 2p} \left[k - \frac{a(1 - s^k)}{1 - s^a} \right], \quad (s \neq 1), \quad d_k = k(a - k), \quad (s = 1),$$

where $s = (1-p)/p$, a is the total stake, k is the gambler's initial stake, and p is the probability that the gambler wins at each play. Display the expected duration d_k against k for a given value of a ($a = 20$ is used in the program) for a selection of values of p.

Find the maximum expected duration for different values of p and plot the results.

Chapter 3: Random Walks

11-3.1. The program simulates a symmetric random walk which starts at position $k = 0$. Display the walk in a graph with joined successive steps in the k (position of walk) versus n (number of steps) plane as illustrated in Figure 3.3. You could try, for example, 1,000 steps.

11-3.2. From the answer to Problem 3.4, calculate the probability that the first return of a symmetric random walk to the origin can be expressed in the form show that

$$f_{2m} = (-1)^{n+1} \binom{\frac{1}{2}}{\frac{1}{2}m}.$$

It has been shown that the mean associated with f_n is infinite. Note that if n is an odd integer, then all f_n are zero. Compute the series

$$s(r) - \sum_{m=1}^{r} 2m f_m$$

for $r = 2, 4, 6, \ldots, 100$, and plot $s(r)$ versus r. Show how $s(r)$ increases with r.

11-3.3. In Section 3.3 it is shown that the probability that a random walk with parameter p is in state x after n steps from the origin is given by

$$v_{n,x} = \binom{n}{\frac{1}{2}(n+x)} p^{\frac{1}{2}(n+x)}(1-p)^{\frac{1}{2}(n-x)},$$

where n and x are either both odd or both even. Define a function of n, x, and p to represent $v_{n,x}$. For specific values of n confirm that the expected value of the final position of the walk is $n(2p-1)$. Again for specific values of n (say 16) and p (say $\frac{2}{3}$), show graphically the probability distribution $v_{n,x}$ in terms of x.

11-3.4. The program simulates a two-dimensional symmetric random walk which starts at the origin. Make sure that odd and even values of the numbers of steps are included (see Problems 3.21 and 3.22). Collect the data of the final positions of n trial walks over m steps. Possible numbers are $n = 2,000$ and $m = 40$. The data can be displayed in a three-dimensional surface plot.

Let the final positions of the walk be the list (i_k, j_k) for $k = 1, 2, 3, \ldots, n$. Calculate the squares $D_k^2 = i_k^2 + j_k^2$, and find the mean of D_k^2. Try this for different m. What would you guess is the mean value of D_k^2 in terms of m?

Chapter 4: Markov Chains

11-4.1. The program computes the eigenvalues and eigenvectors of the transition matrix T given by

$$T = \begin{bmatrix} \frac{1}{4} & \frac{1}{8} & \frac{3}{8} & \frac{1}{4} \\ \frac{1}{3} & \frac{1}{6} & \frac{1}{6} & \frac{1}{3} \\ \frac{1}{3} & \frac{1}{3} & 0 & \frac{1}{3} \\ \frac{1}{3} & 0 & 0 & \frac{2}{3} \end{bmatrix}$$

(also see Problem 4.6(c)). If C is the matrix of eigenvectors and D is the diagonal matrix of eigenvalues, check by a program that $T = CDC^{-1}$. Find a general formula for T^n. For how many decimal places is $\lim_{t \to \infty} T^n$ correct for T^{10}?

11-4.2. Find the eigenvalues and eigenvectors of the stochastic matrix

$$T = \begin{bmatrix} \frac{1}{4} & \frac{1}{2} & \frac{1}{4} \\ \frac{1}{2} & \frac{1}{4} & \frac{1}{4} \\ \frac{1}{4} & \frac{1}{4} & \frac{1}{2} \end{bmatrix}.$$

Construct a formula for T^n, and find $\lim_{n \to \infty} T^n$.

11-4.3. This program simulates a two-state Markov chain with transition matrix

$$T = \begin{bmatrix} 1 - \alpha & \alpha \\ \beta & 1 - \beta \end{bmatrix}.$$

Possible values for the parameters are $\alpha = \frac{1}{2}$ and $\beta = \frac{1}{3}$ for, say, 1,000 steps. Then count the number of times that states E_1 and E_2 occur in the random process. Compare this output with the theoretical stationary distribution

$$\mathbf{p} = \begin{bmatrix} \frac{\beta}{\alpha + \beta} & \frac{\alpha}{\alpha + \beta} \end{bmatrix}.$$

11-4.4. A program is devised to simulate a three-state (or higher state) Markov chain with transition matrix

$$T = \begin{bmatrix} 1 - \alpha_2 - \alpha_3 & \alpha_2 & \alpha_3 \\ \beta_1 & 1 - \beta_1 - \beta_3 & \beta_3 \\ \gamma_1 & \gamma_2 & 1 - \gamma_1 - \gamma_2 \end{bmatrix}.$$

Test the program with the data from Example 4.4 in which

$$T = \begin{bmatrix} \frac{1}{4} & \frac{1}{2} & \frac{1}{4} \\ \frac{1}{2} & \frac{1}{4} & \frac{1}{4} \\ \frac{1}{4} & \frac{1}{4} & \frac{1}{2} \end{bmatrix}.$$

Count the number of times that states E_1, E_2, and E_3 occur in the random process. Compare this output with the theoretical stationary distribution.

11-4.5. A four-state Markov chain with two absorbing states has the transition matrix

$$S = \begin{bmatrix} 1 & 0 & 0 & 0 \\ \frac{3}{4} & 0 & \frac{1}{4} & 0 \\ 0 & \frac{1}{4} & 0 & \frac{3}{4} \\ 0 & 0 & 0 & 1 \end{bmatrix}$$

(see Problem 4.8b). It simulates the chain which starts in state E_2, and determines the proportion of chains which finish in either E_1 or E_2 over, say, 1,000 runs. Check this answer to Problem 4.8(b) by finding $\lim_{t \to \infty} T^n$.

11-4.6. As in Problem 4.8(a), a Markov chain has the transition matrix

$$T = \begin{bmatrix} \frac{1}{4} & \frac{1}{4} & \frac{1}{2} \\ 1 & 0 & 0 \\ \frac{1}{2} & \frac{1}{4} & \frac{1}{4} \end{bmatrix}.$$

Confirm the eigenvalues of T and show that T has just two distinct nontrivial eigenvalues. Find the Jordan decomposition matrix J and a matrix C such that $T = CJC^{-1}$. Input a formula for J^n, find T^n and its limit as $n \to \infty$.

11-4.7. This represents a Markov chain as a **signal flow** of the states $E_1, E_2, \ldots$ against the steps $n = 1, 2, 3, \ldots$. It displays the chain for the three-state process with transition matrix

$$T = \begin{bmatrix} \frac{1}{4} & \frac{1}{4} & \frac{1}{2} \\ 1 & 0 & 0 \\ \frac{1}{2} & \frac{1}{4} & \frac{1}{4} \end{bmatrix}$$

(see Problem 4.8a).

Apply the signal flow display to the six-state chain

$$T = \begin{bmatrix} \frac{1}{4} & \frac{1}{2} & 0 & 0 & 0 & \frac{1}{4} \\ 0 & 0 & 0 & 0 & 0 & 1 \\ 0 & \frac{1}{4} & 0 & \frac{1}{4} & \frac{1}{2} & 0 \\ 0 & 0 & 0 & 0 & 1 & 0 \\ 0 & 0 & 0 & \frac{1}{2} & \frac{1}{2} & 0 \\ 0 & 0 & 0 & \frac{1}{2} & \frac{1}{2} & 0 \\ 0 & 0 & 1 & 0 & 0 & 0 \end{bmatrix}$$

which also appears in Problem 4.11. The flow diagram should reveal the closed subset.

11-4.8. Form an $n \times n$ matrix whose rows are randomly and independently chosen from the uniform discrete distribution of the numbers $1, 2, 3, \ldots, n$. Transform this into a stochastic matrix T by dividing the elements of each row by the sum of the elements in that row. This is a method of generating a class of *positive* stochastic matrices (see Section 4.7a). Find the eigenvalues and eigenvectors of T, T^n, and $\lim_{n \to \infty} T^n$.

Chapter 5: Poisson Processes

11-5.1. In the Poisson process with parameter λ, the probability that the population is of size n at time t is given by $p_n(t) = (\lambda t)^n e^{-\lambda t}/n!$ (see Section 5.2). The program computes the first few probabilities in dimensionless form against λt as in Fig. 5.1.

This also shows that the maximum value of the probability on the graph of $p_n(t)$ lies on the graph of $p_{n-1}(t)$.

11-5.2. Simulate the Geiger counter which is modelled by a Poisson process with parameter λ. The probability of a recording of a hit in any time interval of duration δt is assumed to be $\lambda \delta t$ with multiple recordings negligible (see Section 5.2). Assuming that the readings start from zero, the growth of the readings against time are shown.

Run the program a number of times over the same time interval, and find the mean value of the data. Compare this value with the theoretical value for the mean of a Poisson process.

Chapter 6: Birth and Death Processes

11-6.1. The program simulates a simple birth and death process with birth rate λ and death rate μ. Time interval $(0, T)$ should be divided into increments δt, so that the probability that a birth takes place in any interval of duration δt is $\lambda \delta t$, and similarly for a death (see Section 6.5). Shows graphically how the population changes with time.

11-6.2. The probability generating function $G(s, t)$ for a simple birth and death process with birth and death rates λ and μ, respectively, is given by

$$G(s,t) = \left[\frac{\mu(1-s) - (\mu - \lambda s)e^{-(\lambda-\mu)t}}{\lambda(1-s) - (\mu - \lambda s)e^{-(\lambda-\mu)t}} \right]^{n_0},$$

(see Eqn (6.23). Define the function $G(s, t)$ in *Mathematica*, and find the mean and variance of the population size as functions of t.

11-6.3. In a death process, the death rate is μ and the initial population is n_0. The probability that the population is extinct by time t is given by

$$p_0(t) = (1 - e^{-\mu t})^{n_0}$$

(see Eqn (6.17)). Show that the mean time τ_{n_0} to extinction is given by

$$\tau_{n_0} = \frac{n_0}{\mu} \sum_{k=0}^{n_0-1} \frac{(-1)^k}{(k+1)^2} \binom{n_0 - 1}{k},$$

(see Problem 6.29). The program calculates the dimensionless $\mu\tau_{n_0}$ from the series. Plot a graph of τ_{n_0} versus n_0, say for $n_0 = 1, 2, 3, \ldots, 50$.

Now construct a program to simulate a death process with parameter μ. Calculate a series of extinction times (say, 50?) and compare the mean of these times with the theoretical value from the series.

11-6.4. For a simple birth and death process with equal birth and death rates λ, the probability generating function is

$$G(s,t) = \left[\frac{1 + (\lambda t - 1)(1 - s)}{1 + \lambda t(1 - s)} \right]^{n_0},$$

where n_0 is the initial population size (see Eqn (6.24)). Find the vector function for the probabilities $p_n(t)$ for $n = 0, 1, 2, \ldots, n_1$ as a function of the dimensionless time $\tau = \lambda t$. Show graphically how $p_n(t)$ behaves as τ increases from zero. Possible values are $n_0 = 15$, $n_1 = 30$, and $\tau = 1, 2, 3, 4, 5$.

11-6.5. The partial differential equation for the probability generating function for the birth and death process with birth and death rates λ and μ is given by

$$\frac{\partial G(s,t)}{\partial t} = (\lambda s - \mu)(s - 1)\frac{\partial G(s,t)}{\partial s}.$$

The program solves this equation for $G(s, t)$ subject to the initial population n_0.

Chapter 7: Queues

11-7.1. This program simulates a single-server queue with arrivals having a Poisson distribution with parameter λ and with service times having a negative exponential distribution with parameter μ ($\lambda < \mu$). Possible values are $\lambda = 1$, $\mu = 1.1$ with an elapse time of $t = 2,000$ in steps of duration $\delta t = 0.1$. By counting the number of times queue lengths $0, 1, 2, \ldots$ occur, a distribution for the simulation can be calculated and represented in a bar chart.

The theoretical probability that r persons are in the queue (including the person being served) in the *stationary* process is given by $p_r = (1 - \rho)\rho^r$ where $\rho = \lambda/\mu$. Compare graphically these probabilities with the bar chart.

11-7.2. Consider the solution for the problem of the queue with r servers given in Section 7.4. The probability p_0 that there is no one queueing is given by Eqn (7.11):

$$p_0 = 1 \left/ \left[\sum_{n=0}^{r-1} \frac{\rho^n}{n!} + \frac{\rho^r}{(r - \rho)(r - 1)!} \right] \right. .$$

Represent this by a function of the traffic density $\rho_1 = \rho/r$ and the number of servers r. Plot the probabilities against ρ_1 for $r = 1, 2, 3, 4$ servers. The probability p_n that the queue is of length n is given by

$$p_n = \begin{cases} \rho^n p_0 / n! & n < r \\ \rho^n p_0 / [r^{n-r} r!] & n \geq r. \end{cases}$$

Compose a list of the probabilities when $r = 6$ and $n = 0, 1, 2, \ldots, 20$ for the case $\rho_1 = 0.8$. For what value of n does the probability take its largest value?

Display a surface over the grid $r \times n$ where $r = 1, 2, 3, 4, 5, 6$ and $n = 0, 1, 2, \ldots, 20$ again for the case $\rho_1 = 0.8$. For the same parameter plot the expected queue lengths for $r = 1, 2, \ldots, 20$ excluding those being served.

11-7.3. The fixed service time queue has the probability generating function

$$G(s) = \frac{(1 - \rho)(1 - s)}{1 - se^{\rho(1-s)}}, \quad 0 < \rho < 1,$$

where $\rho = \lambda/\tau$ and τ is the service time. Using *Mathematica* the results in Problem 7.12 for p_0, p_1, p_2 and the expected length of the queue are checked.

11-7.4. In the baulked queue (see Example 7.1) not more than $m \geq 2$ persons are permitted to form a queue. The probability generating function for the queue is (see Problem 7.13) given by

$$G(s) = \frac{(1 - \rho)[1 - (\rho s)^{m+1}]}{(1 - \rho^{m+1})(1 - \rho s)}, \quad \rho \neq 1.$$

This function is defined in *Mathematica*, and a list is obtained which represents the probability distribution over $i = 0, 1, 2, \ldots, m$.

Now compare these results with a simulated baulked queue. Some possible parametric values are

$$\lambda = 1.2; \quad \mu = 1; \quad \text{incremental time step, } \delta t = 0.1; m = 10,$$

run over a time of $t = 2,000$.

Chapter 8: Reliability and Renewal
11-8.1. In Section 8.2, the reliability function is defined as

$$R(t) = \mathbf{P}(T > t) = 1 - F(t),$$

and the failure rate function is given by $r(t) = f(t)/R(t)$. The functions $F(t)$, $R(t)$, and $r(t)$ are derived for the gamma density function $f(t) = \lambda^2 t e^{-\lambda t}$. Plot the functions for $\lambda = 1$ over the interval $0 \leq t \leq 5$.

11-8.2. In Problem 8.3 the failure rate function is given by

$$r(t) = \frac{ct}{a^2 + t^2}, \quad t \geq 0, \quad a > 0, \quad c > 1.$$

Formulae are found for the reliability function $R(t)$ and the density function $f(t)$. All three functions are plotted for $a = 1$ and $c = 2$. What general formula does *Mathematica* give for the mean time to failure in terms of a and c? Find the actual means in the cases (a) $a = 5$, $c = 1.02$, (b) $a = 5$, $c = 2$. Plot the mean time to failure as a surface in terms of a and c for, say, $0.2 < a < 10$ and $1.1 < c < 2$.

11-8.3. Define a reliability function for a process in which n components are in parallel, and each has the same time to failure, which are independent, identically exponential distributions with parameter λ (see Example 8.4). Find the mean time to failure for the cases $n = 1, 2, 3, \ldots, 10$, say. How do you expect this mean to behave as $n \to \infty$?

Define a reliability function for the case of three components in parallel, each having an exponentially distributed failure time, but with different parameters λ_1, λ_2, and λ_3.

Chapter 9: Branching and Other Random Processes

11-9.1. A branching process starts with one individual, and the probability that any individual in any generation produces j descendants is given by $p_j = 1/2^{j+1}$ $(j = 0, 1, 2, \ldots)$. The probability generating function for the n-th generation is given by

$$G_n(s) = \frac{n - (n-1)s}{n + 1 - ns}$$

(see Example 9.1). Find the probability that there are r individuals in the n-th generation. Plot the first few terms in the distributions for the cases $n = 1, 2, 3, 4, 5$. Check the mean population size in the n-th generation.

11-9.2. The branching process in Problem 9.3 has the probability generating function

$$G(s) = a + bs + (1 - a - b)s^2, \quad a > 0, \quad b > 0, \quad a + b < 1.$$

The process starts with one individual. Using a nesting command, the probabilities that there are $0, 1, 2, \ldots, 31, 32$ individuals in the 5th generation, assuming, say, that $a = \frac{1}{4}$ and $b = \frac{1}{3}$, are found. The probabilities are plotted against the number of individuals: why do they oscillate? Also find the probability of extinction for general values of a and b.

11-9.3. In Section 9.5, the gambling problem in which a gambler starts with £1 and bets against the house with an even chance of collecting £2 or losing £1 is explained. If the gambler loses then he or she doubles the bet to £2 with the same evens chances. The gambler continues doubling the bet until he or she wins. It was shown that this eventually guarantees that the gambler wins £1. A program is devised which simulates this martingale, and gives, by taking the mean of a large number of games, an estimate of the expected number of plays until the gambler wins.

11-9.4. This program models the doubling gambling martingale in Section 9.5 to simulate the expected value

$$\mathbf{E}(Z_{n+1} | Z_0, Z_1, Z_2, \ldots, Z_n) = Z_n,$$

where $Z_0 = 1$. The final outputs should take values in the sample space

$$\{-2^n + 2m + 2\}, \quad (m = 0, 1, 2, \ldots, 2^n - 1),$$

for the random variable. A possible case to try is $n = 5$ over 5,000 trials.

11-9.5. This program computes the progress of stochastic epidemic as in Section 9.8. Assume that two individuals in a population size $n = 100$ are infected initially ($t = 0$). The tridiagonal matrix T will need to be defined for $\beta = 1$, $\gamma = 0.5$ and the time-step $\tau = 0.01$. The initial probability vector is

$$\mathbf{s}_0 = \begin{bmatrix} 0 & 0 & 1 & 0 & \cdots & 0 \end{bmatrix}.$$

The calculation $\mathbf{s}_0 T$ is required to determine the random output for $\mathbf{s}(1)$ which will determine $I(\tau)$. This process is repeated in a loop to determine the sequences $\{\mathbf{s}_k\}$ and $\{I(k\tau)\}$ for $k = 1, 2, \ldots 2,000$ if 2,000 steps are chosen. (All data may be varied as long as probability magnitudes are checked.) The graph of the list $I(k\tau)$ versus k can be plotted as in Figure 9.6.

The deterministic simple epidemic is governed by Eqns (9.21) and (9.22):

$$\frac{dZ_S}{dt} = -\frac{\beta}{n} Z_S Z_I + \gamma Z_I,$$

$$\frac{dZ_I}{dt} = \frac{\beta}{n} Z_S Z_I - \gamma Z_I,$$

where Z_S and Z_I represent the susceptible and infective populations. The solution for Z_I is given by (9.24), namely

$$Z_I = \frac{2(\beta - \gamma)n}{2\beta + [(\beta - \gamma)n - 2\beta]e^{-(\beta - \gamma)t}}.$$

Plot the infective curve for $\beta = 1$ and $\gamma = 0.5$, and compare the result with the stochastic curve in Project 11-9.5.

Chapter 10: Brownian Motion: Wiener Process

11-10.1. This program simulates a one-dimensional standard Brownian motion (Wiener process) $X(t)$ over the interval $t \in [0, 1]$ with (say) 1,000 steps. Run several simulations to see how the output can vary.

11-10.2. This program simulates a Brownian motion with drift. Figure 10.2 was drawn with drift 0.3 and volatility 0.3 with 1,000 steps over the time interval $t \in [0, 1]$.

11-10.3. This is a program to determine the time (or time-step) where a standard Brownian motion $X(t)$ first hits a prescribed level $a > 0$. The procedure stops at this time, and plots the process as far as the stopping time.

11-10.4. This program plots five (or any number) of standard Brownian motions with 1,000 steps over $0 \le t \le 10$. It shows how they disperse with respect to the variance $\mathbf{V}(t) = \sqrt{t}$.

11-10.5. This program runs n standard Brownian motions over the same time interval $[0, t]$ and plots the final values of each to show the normal distribution at time t.

11-10.6. An example of geometric Brownian motion together with its trend is illustrated in this program.

Answers and Comments on End-of-Chapter Problems

Answers are given for selected problems.

Chapter 1

1.2. (c) $(A \cap B) \cap C^c$; (e) $[A \cap (B \cup C)^c] \cup [B \cap (A \cup C)^c] \cup [C \cap (A \cup B)^c]$.

1.3. (a) 0.6; (b) 0.1; (c) 0.7.

1.4. The probabilities are $\frac{1}{9}$ and $\frac{5}{9}$.

1.5. The probability is $\frac{5}{12}$.

1.6. The mean value is $\frac{1}{2}$.

1.7. The mean is α and the variance is also α.

1.9. The probability generating function is $ps/(1 - qs)$. The mean is $1/p$ and the variance is q/p^2.

1.10. (a) 11/36; (b) 5/18; (c) 5/36; (d) 1/18; (e) 5/18.

1.11. The expected value of the face values is 7, and the variance is $35/6$.

1.12. There are 216 possible outcomes, the probability that two dice have the same face values and the third is different is $5/12$.

1.13. 17/52.

1.14. Cumulative distribution function is $F(x) = 1 - \frac{1}{2}e^{-(x-a)/a}$.

1.15. $G(s) = ps/[1 - s(1 - p)]$: the mean is $1/p$.

1.16. The expected value is given by

$$\mathbf{E}(Y) = nm/\alpha.$$

1.17. The probabilities are

$$p_0 = \frac{1 - \alpha}{1 + \alpha}, \quad p_n = \frac{2\alpha^n}{(1 + \alpha)^{n+1}} \quad (n = 1, 2, \ldots) \quad \mu = 2\alpha.$$

1.18. The moment generating function is

$$M_X(t) = \frac{1}{b - a} \sum_{n=1}^{\infty} \left(\frac{b^n - a^n}{n!} \right) t^{n-1},$$

and

$$\mathbf{E}(X^n) = \frac{b^{n+1} - a^{n+1}}{(n + 1)(b - a)}.$$

1.20. The means and variances are: (a) μ, $p(1 - p)$; (b) $1/p$, $(1 - p)/p^2$; (c) rq/p; rq/p^2.

1.21. $(0.95)^5 \approx 0.774$ and $(0.9975)^5 \approx 0.9876$.

1.22. (a) 0.000328; (b) 0.1638; (c) 0.01750.

1.25. The chance of choosing 6 numbers correctly is 1 in 13,983,816.

1.26. $\mathbf{P}(\text{vowel}) = 0.3196$.

1.27. $\mathbf{E}(Y|X) = (\frac{4}{3}, \frac{7}{3}, \frac{17}{3})$; $\mathbf{E}[\mathbf{E}(Y|X)] = 39/20$.

1,28. (b) $N(0, 1)$; (c) $N(a\mu + b, a^2\sigma^2)$; (d)$N(0, n)$.

1.29. (c) $\mu = n$; $\sigma^2 = 2n$.

Chapter 2

2.1. (a) Probability of ruin is $u_k \approx 0.132$, and the expected duration is $d_k \approx 409$;
    (c) $u_k = 0.2$, $d_k = 400$.

2.2. The probability $p = 0.498999$ to 6 digits: note that the result is very close to $\frac{1}{2}$.

2.3. (a) $u_k = A + 3^k B$; (b) $A + (B/7^k)$; (c) $u_k = A + (1 + \sqrt{2})^k B + (1 - \sqrt{2})^k C$.

2.4. (a) $u_k = (625 - 5^k)/624$; (c) $u_k = k(10 - k)$.

2.6. The expected duration is $k(a - k)$.

2.7. $d_k = k(a - 2p)/(2p)$.

2.8. The game would be expected to take four times as long.

2.9. The probabilities of the gambler winning in the two cases are 2.23×10^{-4} and 2.06×10^{-6}.

2.10. $p_n = \frac{2}{3} + \frac{1}{3}(-\frac{1}{2})^n$, $n = 1, 2, 3, \ldots$.

2.11. $p_n = [1 - (q - 1)^{n+1}]/(2 - q)$.

2.14. The game extended by

$$\frac{1}{1 + 2p}\left[k - \frac{a(1 - s^k)(s^a - 1 - 2s^k)}{1 - s^{2a}}\right].$$

2.20. The expected duration of the game is about 99 plays.

2.22. (a) $\alpha_k = \frac{1}{2}$; (b) $\alpha_k = (2k - 1)/(4k)$; (c) $\alpha_k = (a + k + 1)/[2(a + k)]$.

2.24. (b) $(e - 1)/e$; (c) $[(n - 1)/n]^{n-1}$.

Chapter 3

3.2. The probability that the walker is at the origin at step 8 is

$$\mathbf{P}(X_8 = 0) = \frac{1}{2^8}\binom{8}{4} \approx 0.273.$$

The probability that it is not the first visit there is approximately 0.234.

3.4. (a)

$$f_n = \begin{cases} (-1)^{\frac{1}{2}n+1}\binom{\frac{1}{2}}{\frac{1}{2}n} & (n \text{ even}) \\ 0 & (n \text{ odd}) \end{cases}.$$

3.5. Treat the problem as a symmetric random walk with a return to the origin: the probability is

$$v_{n,0} = \frac{(2n)!}{2^{2n}n!n!}.$$

3.7. The mean number of steps to the first return is $4pq/|p - q|$.

3.10. The probability is $(2n)!/[2^{2n+1}n!(n + 1)!]$.

3.12. The respective probabilities are $\frac{1}{4}$, $\frac{1}{8}$, and $\frac{1}{64}$.

3.14. The probability that the walk ever visits $x > 0$ is $(1 - |p - q|)/(2q)$.

3.17. (b) The probability that the walk is at O after n steps is

$$
p_j = \begin{cases}
\binom{j}{\frac{1}{2}j}\frac{1}{2^j} + \binom{j}{\frac{1}{2}(j+n)}\frac{1}{2^j} + \binom{j}{\frac{1}{2}(j-n)}\frac{1}{2^j} & (j, n \text{ both even}) \\
0 & (j \text{ odd}, n \text{ even, or } j \text{ even}, n \text{ odd}) \\
\binom{j}{\frac{1}{2}(j+n)}\frac{1}{2^j} + \binom{j}{\frac{1}{2}(j-n)}\frac{1}{2^j} & (j, n \text{ both odd})
\end{cases}
$$

3.18. (a) 0.104; (b) 0.415.

3.19. $p = (1+\sqrt{5})/4$.

3.20. The probability that they are both at the origin at step n is at 0 if n is odd, and, if n is even,

$$
\frac{1}{2^{2n}}\left(\frac{n}{\frac{1}{2}n}\right)^2.
$$

3.21. $p_{2n} \sim 2/(n\pi)$ for large n.

Chapter 4

4.1. $p_{23} = 5/12$; $\mathbf{p}^{(1)} = [\frac{173}{720}, \frac{1}{3}, \frac{307}{720}]$.

4.2. $p_{22}^{(2)} = \frac{13}{36}$; $p_{13}^{(2)} = \frac{13}{48}$; $p_{31}^{(2)} = \frac{17}{48}$.

4.3. $\mathbf{p}^{(3)} = [943, 2513]/3456$; eigenvalues of T are $1, \frac{1}{12}$;

$$
T^n \to \frac{1}{11}\begin{bmatrix} 3 & 8 \\ 3 & 8 \end{bmatrix} \quad \text{as} \quad n \to \infty.
$$

4.5. Eigenvalues are

$$
a+b+c, \quad \frac{1}{2}[2a - b - c \pm i\sqrt{3}|b - c|].
$$

(a) The eigenvalues are $1, \frac{1}{4}, \frac{1}{4}$ with corresponding eigenvectors

$$
\begin{bmatrix} 1 \\ 1 \\ 1 \end{bmatrix}, \quad \begin{bmatrix} -1 \\ 1 \\ 0 \end{bmatrix}, \quad \begin{bmatrix} -1 \\ 0 \\ 1 \end{bmatrix}.
$$

4.6.(a)

$$
T^n = \frac{1}{11}\begin{bmatrix} 4 + 7(-\frac{3}{8})^n & 7 - 7(-\frac{3}{8})^n \\ 4 - (-\frac{3}{8})^n & 7 + (-\frac{3}{8})^n \end{bmatrix}.
$$

(b) The eigenvalues are $-\frac{1}{4}, \frac{1}{4}, 1$.

4.7. 40% of the days are sunny.

4.8. (a) The eigenvalues are $-\frac{1}{4}, -\frac{1}{4}, 1$; (c) The eigenvalues are $\frac{1}{3}, \frac{1}{3}$, and 1. The invariant distribution is $[\frac{1}{4}, 0, \frac{3}{4}]$.

4.9. (a) Eigenvalues are $-\frac{1}{8}, -\frac{1}{8}, 1$.

(b) A matrix C of transposed eigenvectors is given by

$$
C = \begin{bmatrix} 1 & \frac{5}{12} & 1 \\ -4 & -\frac{5}{2} & 1 \\ 1 & 1 & 1 \end{bmatrix}.
$$

(c) The invariant distribution is $[28, 15, 39]/82$.

(d) The eigenvectors are $-\frac{1}{4}, \frac{1}{12}, 1$.

(e) The limiting matrix is

$$
\lim_{n \to \infty} T^n = \begin{bmatrix} 1 & 0 & 0 & 0 \\ \frac{2}{3} & 0 & \frac{1}{3} & 0 \\ 0 & 0 & 1 & 0 \\ \frac{1}{3} & 0 & \frac{2}{3} & 0 \end{bmatrix}.
$$

4.10. $f_1 = 1$; $f_2 = 1$; $f_3 = \frac{1}{16}$; $f_4 = \frac{1}{16}$. Every row of $\lim_{n \to \infty} T^n$ is $\{\frac{2}{3} \quad \frac{1}{3} \quad 0 \quad 0\}$.

4.11. E_4 and E_5 form a closed subset with a invariant distribution $[\frac{1}{3} \quad \frac{2}{3}]$

4.12. All states except E_7 have period 3.

4.15. $\mu_1 = 1 + a + ab + abc$. E_1 is an ergodic state.

4.17. Probability of an item being completed is $(1 - p - q)^2/(1 - q)^2$.

4.18. $\mu_3 = \frac{5}{2}$.

4.23. For example $f_{11}^{(4)} = \frac{5}{48}$.

Chapter 5

5.1. (a) 0.602; (b) 0.235.

5.6. The mean number of calls received by time t is

$$
\mu(t) = at + (b/\omega) \sin \omega t.
$$

5.7. The mean reading at time t is $n_0 + \lambda t$.

5.8. (a) $\mathbf{P}[N(3) = 6] = 0.00353$; (c) $\mathbf{P}[N(3.7) = 4|N(2.1) = 2] = 0.144$; (d) $\mathbf{P}[N(7) - N(3) = 3] = 0.180$.

5.9. The bank should employ 17 operators.

5.10. The expected value of the switch-off times is n/λ.

Chapter 6

6.1. The probability that the original cell has not divided at time t is $e^{\lambda t}$. The variance at time t is $e^{2\lambda t} - e^{\lambda t}$.

6.3. The probability that the population has halved at time t is

$$
p_{\frac{1}{2}n_0}(t) = \binom{n_0}{\frac{1}{2}n_0} [e^{-\mu t}(1 - e^{-\mu t})]^{\frac{1}{2}n_0}.
$$

The required time is $\mu^{-1} \ln 2$.

6.4. (b) $p_0(t) = 0$, $\quad p_n(t) = e^{-\lambda t}(1 - e^{-\lambda t})^{n-1}$, $\quad (n = 1, 2, 3, \ldots)$.

6.5.

$$
p_n(t) = \binom{r}{n} \left(\frac{2}{2+t}\right)^r \left(\frac{t}{2}\right)^n.
$$

The mean is $rt/(2 + t)$.

6.6 (b),(c) The probability of extinction at time t is

$$
G(0, t) = \left[\frac{\mu - \mu e^{-(\lambda - \mu)t}}{\lambda - \mu e^{-(\lambda - \mu)t}}\right]^{n_0} \to \begin{cases} (\mu/\lambda)^{n_0} & \text{if } \lambda > \mu \\ 1 & \text{if } \lambda < \mu. \end{cases}
$$

(d) The variance is

$$
n_0(\lambda + \mu)e^{(\lambda - \mu)t}[e^{(\lambda - \mu)t} - 1]/(\lambda - \mu).
$$

6.7. The probability of ultimate extinction is $e^{-\lambda/\mu}$.

6.18. The mean is

$$
\mu = \frac{\lambda}{\mu}.
$$

6.19. The maximum value of $p_n(t)$ is

$$\binom{n-1}{n_0-1} n_0^{n_0} n^{-n} (n-n_0)^{n-n_0}.$$

6.21. The mean population size at time t is

$$n_0 \exp\left[-\int_0^t \mu(s)ds\right],$$

and

$$\mu(t) = \frac{\alpha}{1+\alpha t}.$$

6.22. Ultimate extinction is certain.

6.23. The probability that the population size is n at time t is given by

$$p_0(t) = \frac{1-\mu e^{-t}}{1+\mu e^{-t}}, \quad p_n(t) = \frac{2\mu^n e^{-nt}}{(1+\mu e^{-t})^{n+1}} \quad (n=1,2,\ldots)$$

and its mean is $2\mu e^{-t}$.

6.30. The mean population size at time t is $e^{(\lambda-\mu)t}$.
The probability generating function is

$$G(s,t) = 1 - e^{-\mu t} + \frac{se^{-(\lambda+\mu)t}}{1-(1-e^{-\lambda t})s}.$$

Chapter 7

7.1. (a) $p_n = 1/2^{n+1}$; (c) $7/8$.

7.3.

$$p_n - 1/\left[1 + \sum_{n=1}^{\infty} \frac{\lambda_0\lambda_1\cdots\lambda_{n-1}}{\mu_1\mu_2\cdots\mu_n}\right].$$

The expected length of the queue is $1/\mu$.

7.4. If $m=3$ and $\rho=1$, then the expected length is $\frac{3}{2}$.

7.5. The mean and variance are respectively $\lambda/(\mu-\lambda)$ and $\lambda\mu/(\mu-\lambda)^2$.

7.6. $\rho \approx 0.74$ and the variance is approximately 10.94.

7.8. 6 telephones should be manned.

7.9. If the expected lengths are $\mathbf{E}_M(N)$ and $\mathbf{E}_D(N)$ respectively, then $\mathbf{E}_D(N) < \mathbf{E}_M(N)$ assuming $0 < \rho < 1$, where $\rho = \lambda/\mu$.

7.10. $0 \le \rho < \sqrt{5} - 1$.

7.11. The required probability generating function is

$$G(s) = \left(\frac{2-\rho}{2+\rho}\right)\left(\frac{2+\rho s}{2-\rho s}\right).$$

7.12. (b) The expected length of the queue is $\rho(1-\frac{1}{2}\rho)/(1-\rho)$ and its variance is

$$\frac{\rho}{12(1-\rho)^2}(12 - 18\rho + 10\rho^2 - \rho^3).$$

(c) $\rho = 3 - \sqrt{5}$.

7.13. The expected length is $\frac{1}{2}m$.

7.14. The expected length is $\rho^2/(1-\rho)$.

7.15. The mean service time is $\frac{10}{3}$ minutes.

7.17. The expected waiting time is $1/(4\mu)$.

7.18. The length of the waiting list becomes $0.89 + O(\varepsilon)$.

7.19. The expected value of the waiting time is

$$\frac{1}{\mu}\left(1 + \frac{p_0 \rho^r}{(r-1)!(r-\rho)^2}\right)'.$$

7.20. The probability that no one is waiting except those being served is

$$1 - \frac{p_0 \rho^{r+1}}{r!(r-\rho)},$$

where p_0 is given by Eqn (7.11).

7.21. For all booths the expected number of cars queueing is

$$\frac{(r-1)\lambda}{\mu(r-1)-\lambda},$$

and the number of extra vehicles queueing is

$$\frac{\lambda^2}{[\mu(r-1)-\lambda][\mu r - \lambda]}.$$

7.22. $\mu = 11\lambda/10$.

Chapter 8

8.1. The reliability function is

$$R(t) = \begin{cases} 1 & t \le t_0 \\ (t_1 - t)/(t_1 - t_0) & t_0 < t < t_1 \\ 0 & t \ge t_1. \end{cases}$$

The expected life of the component is $\frac{1}{2}(t_0 + t_1)$.

8.2. The failure rate function is $r(t) = \lambda^2 t/(1 + \lambda t)$. The mean and variance of the time to failure are respectively $2/\lambda$ and $2/\lambda^2$.

8.3. The probability density function is given by $f(t) = t/(1 + t^2)^{3/2}$.

8.4. The reliability function is

$$R(t) = \begin{cases} e^{-\lambda_1 t^2}, & 0 < t < t_0 \\ e^{-[(\lambda_1 - \lambda_2)t_0(2t - t_0) + \lambda_2 t^2]}, & t > t_0. \end{cases}$$

8.5. The expected time to maintenance is 61.2 hours.

8.6. The expected time to failure is n/α.

8.7. The mean time to failure of the generator is $1/\lambda_f$ and the mean time for the generator to be operational again is $(1/\lambda_f) + (1/\lambda_r)$.

8.8. The relation between the parameters is given by

$$\lambda = -\frac{1}{1000}\ln\left[\frac{0.999 - (1 + 1000\mu)e^{-1000\mu}}{1 - (1 + 1000\mu)e^{-1000\mu}}\right].$$

8.9. 3 components should be carried.

8.10. (a) $\mathbf{P}(T_1 < T_2) = \lambda_1/(\lambda_1 + \lambda_2)$.

(b) $\mathbf{P}(T_1 < T_2) = \lambda_1^2(\lambda_1 + 3\lambda_2)/(\lambda_1 + \lambda_2)^3$.

8.11. $r(t) \to \lambda_1$ if $\lambda_2 > \lambda_1$, and $r(t) \to \lambda_2$ if $\lambda_2 < \lambda_1$.

8.12. $\mathbf{P}(S_3 \le t) = 1 - (1 + \lambda t + \frac{1}{2}t^2\lambda^2)e^{-\lambda t}$.

Chapter 9

9.1. $p_{n,j} = [(2^n - 1)/2^n]^{j-1}$. The mean population size of the n-th generation is 2^n.
9.2. The probability generating function of the second generation is

$$G_2(s) = \frac{(1-p)(1-ps)}{(1-p+p^2)-ps}.$$

The mean population size of the n-th generation is $\mu_n = [p/(1-p)]^n$.
9.3. The maximum possible population size of the n-th generation is 2^n. It is helpful to draw a graph in the (a, b) plane to show the region where extinction is certain.
9.4. The variance of the n-th generation is $\lambda^{n+1}(\lambda^n - 1)/(\lambda - 1)$.
9.5. The expected size of the n-th generation is $\mu_n = [\lambda^n \tanh^n \lambda]/2^n$.
9.7. (a) The probability that the population size of the n-th generation is m is

$$p_{n,m} = \frac{n^{m-2}}{(n+1)^{m+2}}(m + 2n^2 - 1).$$

(b) The probability of extinction by the n-th generation is $[n/(n+1)]^2$.
(c)1.
9.8. $\mu_n = r\mu^n$.
9.9. (b)

$$\mathbf{E}(Z_n) = \begin{cases} (1 - \mu^{n+1})/(1-\mu) & \mu \neq 1 \\ n+1 & \mu = 1. \end{cases}$$

(d) The variance of Z_n is

$$\mathbf{V}(Z_n) = \frac{\sigma^2(1-\mu^n)(1-\mu^{n+1})}{(1-\mu)(1-\mu^2)},$$

if $\mu \neq 1$, or $\frac{1}{2}\sigma^2 n(n+1)$ if $\mu = 1$.
9.10.(a) $\mu_n = \mu^n$ and $\sigma_n^2 = \mu^n(\mu^n - 1)/(\mu - 1)$ if $\mu \neq 1$.
(b) $\mu_n = [p/(1-p)]^n$.
(c)

$$\mu_n = \left[\frac{rp}{1-p}\right]^n, \qquad \sigma_n^2 = \frac{(pr)^n}{(1-p)^{n+1}(rp-1+p)}\left[\left(\frac{rp}{1-p}\right)^n - 1\right],$$

if $\mu \neq 1$.
9.12. (a) $1/[n(n+1)]$; (b) $1/n$.
The mean number of generations to extinction is infinite.
9.13. The mean number of plants in generation n is $(p\lambda)^n$.
9.14. The probability of extinction by the n-th generation is

$$G_n(0) = \frac{(1-p)[p^n - (1-p)^n]}{p^{n+1} - (1-p)^{n+1}}.$$

9.21. $\mathbf{V}(Z_n) = \frac{1}{3}(2^{2n} - 1)$ and $\mathbf{V}[\mathbf{E}(Z_n|Z_0, Z_1, \ldots, Z_{n-1})] = \mathbf{V}(Z_{n-1}) = \frac{1}{3}(2^{2n-2} - 1)$.
9.22. The expected position of the walk after 20 steps is $10(p_1 + p_2 - q_1 - q_2)$.

Chapter 10

10.1. (a) $1 - \Phi(\sqrt{3}/2)$; (b) 0.635.
10.3(a) $F_Z(z) = \Phi(\ln z)/(\sigma\sqrt{t})$, $(z > 0)$.
10.8(a) pdf of $U(t)$ is $e^{-\frac{1}{2}u}/\sqrt{2\pi u}$
    (c) $\mathbf{P}(R(t) \geq r) = e^{-r^2/(2t)}$.

Appendix

Abbreviations

gcd — greatest common divisor.
iid — independent and identically distributed.
mgf —moment generating function.
pgf — probability generating function.
rv — random variable.

Set notation

A, B — sets or events.
$A \cup B$ — union of A and B.
$A \cap B$ — intersection of A and B.
A^c — complement of A.
$B \backslash A$ — B but not A.
S — universal set.
$\emptyset$ —empty set.

Probability notation

N, X, Y — random variables (usually capital letters in the context of probabilities).
$\mathbf{P}(X = x)$ — probability that the random variable $X = x$.
$\mathbf{E}(X)$ — mean or expected value of the random variable X.
$\mathbf{V}(X)$ — variance of the random variable X.
μ — alternative symbol for mean or expected value.
σ^2 — alternative symbol for variance.
Factorial function, $n! = 1 \cdot 2 \cdot 3 \cdots n$; $0! = 1$.
Gamma function, $\Gamma(n) = (n-1)!$, $\qquad\qquad \Gamma(\frac{1}{2}) = \sqrt{\pi}$.
Binomial coefficients:

$$\binom{n}{r} = \frac{n!}{(n-r)!r!}, \qquad n, r \text{ positive integers } n \geq r; \qquad \binom{n}{0} = 1;$$

$$\binom{a}{r} = \frac{a(a-1)\cdots(a-r+1)}{1 \cdot 2 \cdot 3 \cdots r}, \qquad a \text{ any real number.}$$

Power series

Geometric series:

$$1 + s + s^2 + \cdots + s^n = \sum_{j=0}^{n} s^j = \frac{1 - s^{n+1}}{1 - s}, \quad (s \neq 1).$$

$$s + 2s^2 + 3s^3 + \cdots + ns^n = \sum_{j=1}^{n} js^j = \frac{s[1 - (n+1)s^n + ns^{n+1}]}{(1 - s)^2}.$$

Exponential function:

$$e^s = \sum_{n=0}^{\infty} \frac{s^n}{n!} = 1 + s + \frac{s^2}{2!} + \frac{s^3}{3!} + \cdots \quad \text{(for all } s\text{).}$$

Binomial expansions:

$$(1 + s)^n = \sum_{r=0}^{n} \binom{n}{r} s^r$$

$$= 1 + ns + \frac{n(n-1)}{2!} s^2 + \cdots + ns^{n-1} + s^n, \quad (n \text{ is a positive integer})$$

$$(1 + s)^a = \sum_{r=0}^{\infty} \binom{a}{r} s^r$$

$$= 1 + as + \frac{a(a-1)}{2!} s^2 + \frac{a(a-1)(a-2)}{3!} s^3 + \cdots$$

$$(|s| < 1, a \text{ any real number}).$$

Probability generating function

$$G(s, t) = \sum_{n=0}^{\infty} p_n(t) s^n,$$

$$\text{mean} = \boldsymbol{\mu}(t) = G_s(1, t) = \sum_{n=0}^{\infty} n p_n(t),$$

$$\text{variance} = \sigma^2(t) = G_{ss}(1, t) + G_s(1, t) - G_s(1, t)^2.$$

Order notation

'Big O' notation: $f(t) = O(g(t))$ as $t \to \infty$ means $f(t)/g(t) < K$ for some constants K and t_0, and all $t > t_0$.

'Little o' notation: $f(t) = o(g(t))$ as $t \to \infty$ means $\lim_{t \to \infty} f(t)/g(t) = 0$. In particular $f(t) = o(1)$ means $\lim_{t \to \infty} f(t) = 0$.

Integrals

$$\int_0^\tau e^{-\lambda t} dt = \frac{1}{\lambda}(1 - e^{-\lambda \tau}), \qquad \int_0^\infty e^{-\lambda t} dt = \frac{1}{\lambda}.$$

$$\int_0^\tau te^{-\lambda t}dt = \frac{1}{\lambda^2}[1-(1+\lambda t)e^{-\lambda t}], \qquad \int_0^\infty te^{-\lambda t}dt = \frac{1}{\lambda^2}.$$

$$\int_0^\infty t^n e^{-\lambda t}dt = \frac{n!}{\lambda^{n+1}} \qquad \int_0^\infty e^{-\lambda t^2}dt = \frac{1}{2}\sqrt{\frac{\pi}{\lambda}}.$$

Matrix algebra

Transition (stochastic) matrix $(m \times m)$:

$$T = [p_{ij}] = \begin{bmatrix} p_{11} & p_{12} & \cdots & p_{1m} \\ p_{21} & p_{22} & \cdots & p_{2m} \\ \vdots & \vdots & \ddots & \vdots \\ p_{m1} & p_{m2} & \cdots & p_{mm} \end{bmatrix}, \qquad (0 \le p_{ij} \le 1 \text{ for all } i,j)$$

where

$$\sum_{j=1}^m p_{ij} = 1.$$

Matrix product of $A = [a_{ij}]$, $B = [b_{ij}]$, both $m \times m$:

$$AB = [a_{ij}][b_{ij}] = \left[\sum_{k-1}^m a_{ik}b_{kj} \right].$$

Diagonal matrix $(m \times m)$ with diagonal elements $\lambda_1, \lambda_2, \ldots, \lambda_m$:

$$D = \begin{bmatrix} \lambda_1 & 0 & \cdots & 0 \\ 0 & \lambda_2 & \cdots & 0 \\ \vdots & \vdots & \ddots & \vdots \\ 0 & 0 & \cdots & \lambda_m \end{bmatrix}.$$

$m \times m$ identity matrix

$$I_m = \begin{bmatrix} 1 & 0 & \cdots & 0 \\ 0 & 1 & \cdots & 0 \\ \vdots & \vdots & \ddots & \vdots \\ 0 & 0 & \cdots & 1 \end{bmatrix}.$$

For the matrix A, A^t denotes the transpose of A in which the rows and columns are interchanged.

Characteristic equation for the eigenvalues $\lambda_1, \lambda_2, \ldots, \lambda_m$ of the $m \times m$ matrix T:

$$\det(T - \lambda I_m) \quad \text{or} \quad |T - \lambda I_m| = 0.$$

The eigenvectors $\mathbf{r}_1, \mathbf{r}_2, \ldots, \mathbf{r}_m$ satisfy:

$$[T - \lambda_i I_m]\mathbf{r}_i = \mathbf{0}, \qquad (i = 1, 2, \ldots, m).$$

Probability distributions

Discrete distributions:

Distribution	$\mathbf{P}(N = n)$	Mean	Variance
Bernoulli	$p^n(1-p)^{1-n}$ $(n = 0, 1)$	p	$p(1-p)$
Binomial	$\binom{r}{n} p^n(1-p)^{r-n}$ $(n = 0, 1, 2, \ldots, r)$	rp	$rp(1-p)$
Geometric	$(1-p)^{n-1}p$ $(n = 1, 2, \ldots)$	$1/p$	$(1-p)/p$
Negative binomial	$\binom{n-1}{r-1} p^n(1-p)^n$ $(n = r, r+1, \ldots)$	r/p	$r(1-p)/p^2$
Poisson	$e^{-\alpha}\alpha^n/n!$, $(n = 0, 1, 2, \ldots)$	α	α
Uniform	$1/k$, $(n = r, r+1, \ldots, r+k-1)$	$\frac{1}{2}(k + 2r - 1)$	$\frac{1}{12}(k^2 - 1)$

Continuous distributions

Distribution	Density, $f(x)$	Mean	Variance
Exponential	$\alpha e^{-\alpha x}$ $(x \geq 0)$	$1/\alpha$	$1/\alpha^2$
Normal, $N(\mu, \sigma^2)$	$\frac{1}{\sigma\sqrt{2\pi}} \exp\left[-\frac{(x-\mu)^2}{2\sigma^2}\right]$ $-\infty < x < \infty$	μ	σ^2
Gamma	$\frac{\alpha^n}{\Gamma(n)} x^{n-1} e^{-\alpha x}$, $(x \geq 0)$	n/α	n/α^2
Uniform	$\begin{cases} 1/(b-a) & a \leq x \leq b \\ 0 & \text{for all other } x \end{cases}$	$\frac{1}{2}(a+b)$	$\frac{1}{12}(b-a)^2$
Weibull	$\alpha\beta x^{\beta-1} e^{-\alpha x^\beta}$, $(x \geq 0)$	$\alpha^{-1/\beta}\Gamma(\beta^{-1} + 1)$	$\alpha^{-2/\beta}[\Gamma(2\beta^{-1} + 1) - \Gamma(\beta^{-1} + 1)^2]$
chi-squared χ_n^2	$\begin{cases} 0 \\ \frac{1}{2\Gamma(\frac{1}{2})}[\frac{1}{2}x]^{\frac{1}{2}n-1} e^{-\frac{1}{2}x} \end{cases}$	n	$2n$

References and Further Reading

Addison, P.S. (1997): *Fractals and Chaos*, Institute of Physics Publishing, Bristol.

Allen, L.J.S. (2008) Ch. 3: An Introduction to Stochastic Epidemic Models in (*Mathematical Epidemiology*, edited by F. Brauer, P. van der Driesshe, J. Wu, Springer, Berlin).

Bailey, N.T.J. (1964): *The Elements of Stochastic Processes*, John Wiley, New York.

Beichelt, F. (2006): *Stochastic Processes in Science, Engineering and Finance*, Chapman & Hall, Boca Raton.

Blake, I.F. (1979): *An Introduction to Applied Probability*, John Wiley, New York.

Cox, D.R., Donnelly, C.A., Bourne, F.J., Gettinby, G., McInerney, J.P., Morrison, W.I., and Woodroffe, R. (2005): Simple model for tuberculosis in cattle and badgers. *Pro. Nat. Acad. Sci.* 102:17588–17593.

Einstein, A. (1956): *Investigations on the Theory of Brownian Motion*, Dover Publications (translated by A.D. Cowper).

Feller, W. (1968): *An Introduction to Probability Theory and Its Applications*, 3rd edition, John Wiley, New York.

Ferguson, N.M., Donnelly, C.A., and Anderson, R.M. (2001): The foot-and-mouth epidemic in Great Britain: pattern of spread and impact of interventions. *Science* 292:1155–1160.

Grimmett, G.R. and Stirzaker, D.R. (2005): *Probability and Random Processes*, Clarendon Press, Oxford.

Grimmett, G.R. and Welsh, D.J.A. (1986): *Probability: An Introduction*, Clarendon Press, Oxford.

Jones, P.W. and Smith, P. (1987): *Topics in Applied Probability*, Keele Mathematical Education Publications, Keele University.

Jordan, D.W. and Smith, P. (2008): *Mathematical Techniques*, 4th ed, Oxford University Press.

Larsen, R.J. and Marx, M.L. (2012): Introduction to Mathematical Statistics and Its Applications, 5th edition, Pearson.

Lawler, G.F. (2006): *Introduction to Stochastic Processes*, 2nd edition, Chapman & Hall/CRC Press, New York.

Luenberger, D.G. (1979): *Introduction to Dynamic Systems*, John Wiley, New York.

Meyer, P.L. (1970): *Introductory Probability and Statistical Applications*, 2nd edition, Addison-Wesley, Reading, MA.

Open University (1988): *Applications of Probability*, Course Units for M343, Open University Press, Milton Keynes.

Ross, S. (1976): *A First Course in Probability*, Macmillan, New York.

Song, S. and Song, J. (2013): A Note on the History of the Gambler's Ruin Problem, *Comm. Stat. Appl. Methods*, 20: 157–168.

Stirzaker, D. (2005): *Stochastic Processes and Models*, Oxford University Press.

Tuckwell, H.C. (1995): Elementary Applications of Probability Theory, Chapman & Hall, London.

Ugarte, M.D., Militino, A. F., and Arnholt, A. (2008): *Probability and Statistics with R*, CRC Press, Boca Raton.

Wolfram, S. (1996): *The Mathematica Book*, 3rd edition, Cambridge University Press.

Ziplin, E.F., Jennelle, C.S., and Cooch, E.G. (2010): A primer on the application of Markov chains to the study of wildlife disease dynamics. *Methods in Ecology & Evolution* 1: 192–198.

Index

Printed and bound by CPI Group (UK) Ltd, Croydon, CR0 4YY

24/10/2024

01778284-0005